Essays in Biochemistry

volume 27

Essays in Biochemistry

edited by K.F. Tipton

PORTLAND PRESS, London and Chapel Hill

Essays in Biochemistry is published by Portland Press Ltd
on behalf of the Biochemical Society

Portland Press Ltd
59 Portland Place
London W1N 3AJ
U.K.

In North America orders should be sent to:

Portland Press Inc.
P.O. Box 2191
Chapel Hill, NC 27515-2191
U.S.A.

British Library Cataloguing-in-Publication Data
A catalogue record for this book is available from the British Library

ISBN   1-85578-015-1

ISSN   0071-1365

Typeset by Portland Press Ltd and printed in
Great Britain by the University Press, Cambridge

Contents

Contents (continued)

Contents (continued)

The authors

Graham Kemp is a lecturer in the newly formed School of Biological and Medical Sciences at the University of St. Andrews. He graduated from the erstwhile Biochemistry Department at St. Andrews and carried out postgraduate research there before taking up postdoctoral positions in Toronto and Bern. He returned to St. Andrews as a lecturer in biochemistry in 1974. **Ailsa Webster** graduated with an honours B.Sc. in biochemistry from the University of St. Andrews and has just completed her Ph.D. thesis in the Department of Biochemistry on the subject of viral proteases. **Willie Russell** has been Professor of Biochemistry at the University of St. Andrews since 1984 having been a member of the M.R.C. Scientific Staff at NIMR, Mill Hill, London from 1964, latterly as Head of the Division of Virology. He obtained an honours degree in chemistry and a Ph.D. from the University of Glasgow and joined the Institute of Virology there at its inception in 1959.

Michael F. Scully is Senior Lecturer in the Thrombosis Research Institute (affiliated to the National Heart and Lung Institute and King's College School of Medicine and Dentistry, University of London) and is a member of the editorial board of *Thrombosis Research*. He is a graduate and doctorate in Biochemistry from the University of Wales, College of Cardiff. His published research is principally upon the mechanism of action of heparin and heparin-like compounds in the control of thrombin and factor Xa. At present, he is investigating the properties of tissue factor pathway inhibitor and the cell biology of plasminogen activation on the cell surface. He is also undertaking collaborative research into the design of synthetic inhibitors of coagulation proteinases. He has co-edited three books: *Chromogenic Peptide Substrates* (1979), *Mechanisms of Stimulus Response Coupling in Platelets* (1985) and *Design of Synthetic Inhibitors of Thrombin* (1992).

Fergus Doherty is a Lecturer in the Department of Biochemistry, University of Nottingham. Current research projects include investigations into the role of ubiquitin-mediated proteolysis in the maturational dependent loss of organelles and proteins from the reticulocyte and the role of ubiquitin and stress proteins in terminal differentiation of the keratinocyte. **R. John Mayer** is Professor of Molecular and Cell Biology, in the Department of Biochemistry, University of Nottingham. Professor Mayer is very much involved in investigating the role of ubiquitin in human disease and in viral infection. He is also investigating the role of ubiquitin and lysosomal proteolysis.

David J.S. Hulmes is a Senior Lecturer and Wellcome Research Leave Fellow in the Department of Biochemistry, University of Edinburgh. After graduating with a B.Sc. (Wales) in physics and a D.Phil. (Oxon) in molecular biophysics, his research career, in the European Molecular Biology Laboratory (Grenoble), Massachusetts General Hospital/Harvard Medical School (Boston), University of Manchester, Rutgers Medical School (Piscataway) and Thomas Jefferson University (Philadelphia), has focused on the structure and assembly of the extracellular matrix. Major achievements have been the elucidation of the molecular packing in collagen fibrils and the role of procollagen processing in fibril assembly. Current research interests include cell–matrix interactions, corneal morphogenesis, lysyl oxidase, emphysema and osteoarthritis.

Leona Samson graduated from Aberdeen University with a B.Sc. in biochemistry in 1974 and from University College, London University (and the Imperial Cancer Research Fund) with a Ph.D. in Molecular Biology in 1978. In 1983, after 5 years of postdoctoral training at the University of California in San Francisco and the University of California in Berkeley, she joined the Molecular and Cellular Toxicology Department at the Harvard School of Public Health as an Assistant Professor and since 1989 has been an Associate Professor in the same department.

Tomoh Masaki, Professor of Pharmacology in the Faculty of Medicine at Kyoto University, was born in Tokyo in 1934. Following graduation from the Tokyo University School of Medicine, he entered Professor Setsuro Ebashi's laboratory at Tokyo University where he started his work on the pharmacology and biochemistry of muscle. He moved to the University of Tsukuba as a Professor of Pharmacology in 1975, and at that time his research focus changed from striated muscle to smooth muscle. He moved to his present position at the Kyoto University School of Medicine in 1991. **Masashi Yanagisawa**, born in Tokyo in 1960, is Assistant Professor of Pharmacology at Kyoto University Faculty of Medicine. Graduation from the University of Tsukuba School of Medicine in 1985 was followed by graduate research in the same institution under the supervision of Professor Masaki, initially on molecular cloning of smooth-muscle and non-muscle myosins, and then on the project that led to the discovery of endothelins in 1987. The award of a Ph.D. in 1988 led to appointment as Assistant Professor of Pharmacology at Tsukuba in 1989, followed by the move to Kyoto in 1991.

Richard Rodnight has spent the main part of his career at the Institute of Psychiatry, London University, where he was Reader and then Professor of Neurochemistry in the Department of Biochemistry. He has published widely on various aspects of protein phosphorylation in the brain. On retirement in 1987 he moved to Brazil where he is presently a Visiting Professor in the Department of Biochemistry of the Federal University of Rio Grande do Sul in Porto Alegre. **Susana Wofchuk** is a Lecturer in the same Department of Biochemistry in Brazil. Her research interests have included aspects of amino acid metabolism and currently protein phosphorylation in the nervous system.

Susan Greenfield initially studied psychology at Oxford before concentrating more on the neurochemistry of the mammalian brain. She originally worked in the laboratory of David Smith, on the release of soluble cholinesterase into cerebrospinal fluid. This research gave her an interest in novel mechanisms of neuronal communication: she worked with Jaques Glowinski in Paris on dendritic release of neuroactive substances, and subsequently with Rodolfo Llinas in New York studying dendritic calcium potentials. Her current research at Oxford involves a multidisciplinary approach to the physiology and pathology of the neurons that degenerate in Parkinson's disease. She is currently a Lecturer in Synaptic Pharmacology and Fellow of Lincoln College, Oxford. Susan Greenfield has a keen extramural interest in the philosophy of mind and has edited a book (with Colin Blakemore) on the interface between philosophy and neuroscience, *Mindwaves*, Basil Blackwell, Oxford, 1987.

William S. McIntire is a Research Chemist in the Molecular Biology Division at the U. S. Department of Veterans Affairs Medical Center, and an Associate Research Biochemist in the Department of Biochemistry and Biophysics and the Department of Anesthesia at the University of California, San Francisco. He received a B.S. degree in chemistry from the University of Illinois, Chicago Circle, and his M.S. and Ph.D. in biochemistry from the University of California, Berkeley. He is author of over 50 papers and chapters. His current research interests focus on the structure, mechanism and biosynthesis of oxidoreductases containing covalently-bound riboflavin and quinone prosthetic groups.

Steve Rawsthorne graduated from the University of Manchester in 1978 with a degree in biochemistry. He followed his Ph.D. studies at the University of Reading on the biochemistry and physiology of the legume/*Rhizobium* symbiosis with post-doctoral studies at the Boyce Thompson Institute, Cornell University, NY, U.S.A. There he investigated mitochondrial respiration under *in vitro* conditions which simulated the very low oxygen environment found in legume root nodules. Returning to the U.K., he started work at the John Innes Institute, Norwich, as a postdoctoral worker and is now a project leader in the Cambridge Laboratory at the same site. His group's work at Norwich has established the biochemistry behind C_3-C_4 intermediate photosynthesis, and now encompasses biochemical, genetical, molecular biological and physiological approaches.

Jeffrey Newman has been involved in research and development of biosensors since he joined the biosensors department at Unilever in 1985. He moved to Bristol Polytechnic in 1987 to study for a Ph.D. in potentiometric microbial biosensors. He is presently employed by the Biotechnology Centre at Cranfield Institute of Technology where his current interests are advanced sensor manufacturing technologies and environmental biosensors. **Anthony Turner** has developed a number of biosensors during his 10 years at Cranfield Institute of Technology and produced over 200 publications and patents in the area. He is an editor of *Biosensors: Fundamentals and Applications* (Oxford), the series *Advances in Biosensors* (SAI) and the journal *Biosensors and Bioelectronics*. He is a well-known speaker at international conferences worldwide. He balances his academic research with an awareness of commercial problems and opportunities gained by consulting for a large number of companies, national and international organizations.

Henry (Hal) Dixon, originally from Dublin, is a lecturer at the University of Cambridge, where he did his first degree and Ph.D. His work on the chemistry of corticotropin led him to transamination for the specific modification of N-terminal residues in native proteins and their stepwise removal. His interest in transamination was broadened by a sabbatical year in the late Professor Braunstein's laboratory in Moscow. His other main work is on phosphonic and arsonic acids as analogues of natural phosphates. When he was asked to lecture on pH, he found many of the concepts difficult, so he was forced to examine how its presentation for enzymology could be clarified.

Abbreviations

The abbreviations and conventions used in *Essays in Biochemistry* generally follow those recommended for use in the *Biochemical Journal* [see *Biochem. J.* (1992) **281**, 1–19]. Other abbreviations used in this volume are:

ANP	atrial natriuretic peptide
cDNA	complementary DNA
EDRF	endothelium-derived relaxing factor
eiF	eukaryotic initiation factor
Gla	γ-carboxyglutamic acid
HIV	human immunodeficiency virus
MPDP	1-methyl-4-phenyl-2,3-dihydropyridine
MPP$^+$	1-methyl-4-pyridinium ion
MPTP	1-methyl-4-phenyl-1,2,3,6-tetrahydropyridine
NMDA	*N*-methyl-D-aspartate
PAI	plasminogen activator inhibitor
PQQ	pyrolloquinoline quinone
RuBP	ribulose 1,5-bisphosphate
TQ	topa quinone
TTQ	tryptophan tryptophylquinone
UCRP	ubiquitin cross-reactive protein

The following abbreviations are used for the genetically encoded amino acids:

Alanine	Ala	A
Arginine	Arg	R
Asparagine	Asn	N
Aspartic acid	Asp	D
Aspartic acid or asparagine	Asx	B
Cysteine	Cys	C
Glutamine	Gln	Q
Glutamic acid	Glu	E
Glutamic acid or glutamine	Glx	Z
Glycine	Gly	G
Histidine	His	H
Isoleucine	Ile	I

Leucine	Leu	L
Lysine	Lys	K
Methionine	Met	M
Phenylalanine	Phe	F
Proline	Pro	P
Serine	Ser	S
Threonine	Thr	T
Tryptophan	Trp	W
Tyrosine	Tyr	Y
Unknown or "other"	Xaa	X
Valine	Val	V

The following symbols are used for nucleotides in nucleic acid sequences (both DNA and RNA; note that the symbol T is used at all positions where U might appear in the RNA):

G	guanine
A	adenine
T	thymine
C	cytosine
N	"any" or "unknown"

Proteolysis is a key process in virus replication

Graham Kemp, Ailsa Webster and W.C. Russell

Department of Biochemistry and Microbiology, School of Biological and Medical Sciences, University of St. Andrews, St. Andrews, Fife KY16 9AL, Scotland, U.K.

Viruses by their nature are not capable of independent existence. They propagate themselves within the environment of a host cell which they invade and colonize, in the process subverting many of the host cell functions for their own purposes of self-replication and assembly. They travel light, carrying within their genome only the information required for the synthesis of their own characteristic coat or capsid proteins plus a few specific components and enzymes essential for the takeover and control of the host cell. The fact that in many viruses of diverse types one of the virus-coded enzymes is a protease attests to the key role that proteolysis plays in viral replication.

Following a brief introduction to the nature of viruses this essay will describe the various functions that proteases carry out in virus replication and discuss the increasing evidence that suggests viruses have evolved proteolytic mechanisms distinct from those of the host cell and developed different strategies for their control. Finally we will examine how the unusual nature of viral proteases, combined with their frequently exquisite specificity and essential role in replication, are being utilized to develop a new generation of antiviral agents.

THE NATURE OF VIRUSES

Infectious virus particles, or virions, are simple structures consisting essentially of two parts — a genome surrounded by a protein coat or capsid. In some instances the capsid is, in turn, surrounded by a membranous envelope. The capsid and envelope, as well as protecting the genome, have important roles in facilitating entry

Table 1. Proteases involved in viral replication
This is a partial list of the proteases involved in viral replication, giving their source (virus-coded unles
otherwise stated), putative classification and presumed function. Key: TL-Cys, trypsin-like cysteine
proteinase; PL-Cys, papain-like cysteine protease.

Family	Genus	Virus	Genome	Protease	Class	Functions
Picornaviridae	Enterovirus	Poliovirus	RNA	3C	TL-Cys	Polyprotein processing
	Rhinovirus	Rhinovirus	RNA	and		and inhibition of host
				2A	TL-Cys	systems
				?	?	Maturation
	Aphthovirus	Foot and	RNA	3C	TL-Cys	Polyprotein processing
		mouth disease		and		and inhibition of host
		virus		L	PL-Cys	systems
	Cardiovirus	Encephalo-	RNA	?	?	Maturation
		myocarditis				
		virus				
Picornaviridae-like plant viruses	Comovirus	Cow pea mosaic virus	RNA	24K	TL-Cys	Polyprotein processing
	Nepovirus	Grapevine fan leaf virus	RNA	VP7	TL-Cys	Polyprotein processing
	Potyvirus	Tobacco etch virus	RNA	NIa	TL-Cys	Polyprotein processing
				HCPro	PL-Cys	Autocatalytic release
				P1	Serine?	Autocatalytic release
Togaviridae	Alphavirus	Sindbis virus	RNA	nsP2	PL-Cys	Polyprotein processing
				Capsid protein	Serine	Autocatalytic release
				Cellular		Envelope maturation
	Rubivirus	Rubella virus	RNA	M-Pro	PL-Cys	Polyprotein processing
	Flavivirus	Yellow fever virus	RNA	NS3	Serine	Polyprotein processing
	Pestivirus	Bovine viral diarrhoea virus	RNA	p20	?	Autocatalytic release
				p80	Serine	Polyprotein processing
				Cellular		Envelope maturation
Retroviridae		Human immuno-deficiency virus	RNA	p11	Aspartic	Polyprotein processing
				Cellular		Envelope maturation
Coronaviridae		Murine hepatitis virus	RNA	L-Pro	PL-Cys	Polyprotein procesing
Adenoviridae	Mastadenovirus	Adenovirus	DNA	23K	Cysteine	Maturation
Poxviridae	Orthopoxvirus	Vaccinia virus	DNA	?	?	?
Herpesviridae	Herpesvirus	Herpes simplex virus	DNA	UL26?	?	?
Myoviridae		T4 phage	DNA	Gp21	Serine	Maturation
Orthomyxoviridae		Influenza virus	RNA	Cellular		Envelope maturation
Paramyxoviridae	Morbillivirus	Measles virus	RNA	Cellular		Envelope maturation

into the host cell by interacting with cell-surface receptors or by fusing with cell membranes.

Because viruses are economical with their utilization of genetic information they exhibit few morphological forms. The capsid structures of the majority are either icosahedral or helical, constructed by the regular packing of a limited number of protein molecules. Exceptions to this are the larger bacteriophages, such as T4, which have multisection coats and the more complex animal viruses such as vaccinia which lack a clearly identifiable capsid, having instead several layers surrounding their genome. The membranous envelope sometimes seen enclosing the capsid consists of viral (glyco)proteins embedded in a lipid layer derived from the host cell.

The genome can be either DNA or RNA, in single- or double-stranded form, varying in length from 3 to 300 kilobases. It is packaged within the capsid along with basic proteins or polyamines, in some cases accompanied by a limited number of enzymes.

VIRAL REPLICATION STRATEGIES

Once inside the cell viruses depend ultimately on host cell mechanisms, but the precise replication strategy they adopt is dictated by the nature of their genetic material. Positive-strand RNA viruses can utilize their RNA directly as the message, translating it as a single large protein known as a polyprotein. Full expression of the various structural proteins and enzymes contained within the polyprotein requires controlled and directed proteolysis. Negative-strand RNA viruses and DNA viruses normally produce monocistronic messages and therefore do not require proteases for polyprotein cleavage, but nonetheless, proteolysis is an important process in the maturation of these viruses[1]. Very often the progress of virus replication is facilitated by the direct suppression of host cell protein and nucleic acid synthesis.

The best studied family of positive-strand RNA viruses is the Picornaviridae which include the viruses responsible for poliomyelitis, many varieties of the common cold, and foot and mouth disease of cattle. A number of plant viruses are similar to the Picornaviruses[2] (Table 1), although some have two separate RNA molecules, one coding for a polyprotein containing the structural proteins and the other responsible for the replicative enzymes. In the comovirus, cow pea mosaic virus, the RNA molecules coding for structural proteins are packaged in separate virions from those coding for non-structural proteins, a productive infection requiring that both virions enter the same cell.

Retroviruses, such as the human immunodeficiency virus (HIV) use a reverse transcriptase to produce DNA from their RNA genome and then incorporate this into the host cell DNA[3]. Transcription from this produces a number of subgenomic RNA molecules which are translated as multidomain polyproteins.

Translation of the genome as a single large polyprotein presents an immediate problem to the virus. All the proteins will be produced in equimolar amounts, although structural proteins will be utilized in stoichiometric amounts while enzymes are required only in catalytic quantities. Like the comoviruses, the togaviruses use separate RNA molecules, subject to different transcriptional and translational control, for structural and non-structural proteins, while some retroviruses have an additional strategy for addressing this problem at the level of replication (Figure 1). While there is some

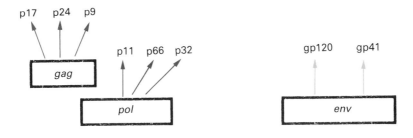

Figure 1. Genetic organization of the HIV genome
The positions of the three polyproteins in the proviral DNA are indicated along with the protein products of each. The protease monomer is p11. *gag* and *pol* are cleaved by the viral protease, *env* by a cellular protease.

variation between different retroviruses, in general their genome consists of three major sections, *gag*, *pol* and *env*, containing the genes for the structural proteins, enzymes and envelope proteins respectively. The section coding for the *gag* poly-protein is at the 5′ end, either in a separate reading frame or with a stop codon separating it from the others. The *gag* polyprotein is therefore always expressed, and occasional read-through of the stop codon (murine leukaemia virus), or a frameshift event (HIV), ensures that smaller amounts of the non-structural proteins are produced. The *env* gene is separately transcribed and translated as a polyprotein. As will be discussed later, controlled proteolysis is an important additional factor in matching expression levels to the requirements of the virus.

THE ROLE OF PROTEOLYSIS IN VIRAL REPLICATION

Not all viruses that depend on proteolysis code for their own protease, many making use of the proteases that abound in the host cell. Viruses that code for their own protease have some selective advantage in that the range of cells in which they can replicate is not limited to those possessing proteases with the appropriate specificity. The *quid pro quo* for this, however, is that the viral protease must have evolved to avoid the wide range of protease inhibitors and control mechanisms that exist in most cells. As a consequence, they would be expected to be unusual molecules — and indeed as more virus-coded proteases are identified it is becoming apparent that many have primary structures unlike those of any non-viral proteases and reaction mechanisms with significant differences from those characterized for enzymes from higher organisms. As has been pointed out in recent reviews of viral proteases[1,4,5] these distinctive mechanisms, developed of necessity, may well represent the Achilles heel of many viruses. Inhibitors developed against viral proteases are likely to have little effect on their host cell counterparts and therefore present a promising approach to antiviral agents.

Polyprotein cleavage presents perhaps the most obvious role for proteolysis in the virus replication cycle, but there are others just as vital. Once the viral capsid has been assembled and the genome inserted, many virions require cleavage of one or more capsid proteins as a final step. Whether this cleavage is required, as it were, to lock the door, or perhaps more likely to weaken the capsid structure to ensure release of the genome following penetration of the host cell, is unclear. What is known, however, is that this final cleavage, or virion maturation as it is known, is an absolute requirement for the development of infectivity in a wide range of virus families (Table 1). Whether proteolysis also plays a direct role in the uncoating process following cell entry is not known but seems a distinct possibility. In common with many higher organisms, viruses also use proteases for activation of their own proteins which are produced as inactive precursors, while some take this a stage further and use proteases to inactivate some host cell systems.

Polyprotein cleavage in picornaviruses

The animal picornaviruses, which translate almost their entire genome as a single polyprotein, represent one extreme of the polyprotein phenomenon, as replication strategy has no direct influence on the relative levels of the individual proteins eventually produced. Thus they present a good starting point to explore this aspect of viral proteolysis. The four animal picornavirus genera are listed in Table 1 along with the plant viruses that can be considered to belong to the Picornaviridae family. The well-characterized proteases of poliovirus will be used to illustrate how proteolysis can be controlled and directed to circumvent the problems associated with a replication strategy based on polyprotein synthesis.

Poliovirus codes for two proteases designated 2A and 3C, the genome locations and cleavage points of which are shown in Figure 2. An early event following poly-

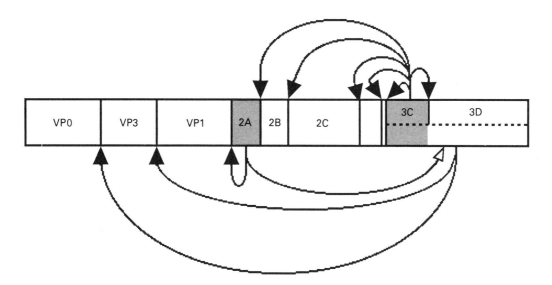

Figure 2. Poliovirus polyprotein
Schematic diagram of the polyprotein produced by poliovirus showing the location of the cleavages carried out by the 2A, 3C and 3CD proteases. The open arrowhead indicates the non-essential cleavage in the polymerase domain carried out by 2A.

protein synthesis, which occurs before the protein leaves the ribosome, is the auto-catalytic cleavage that 2A carries out at its own N-terminus. This serves to separate the structural proteins of the poliovirus from the rest and may be the only essential proteolytic cleavage carried out by 2A. There is a cleavage within the RNA polymerase (3D) which is catalysed by 2A, but this is not obligatory for successful replication[5]. As will be discussed later, 2A also has an important role in helping to switch off host cell DNA synthesis.

All but one of the remaining cleavages necessary for the complete dissection of the polyprotein are carried out by the 3C protease, some of them while it is part of the larger intermediate 3CD protein. Interestingly, there is a distinct division of labour in that the protein released from the N-terminal part of the polyprotein by 2A is cleaved by 3CD to produce the capsid proteins. The replication enzymes are released from the C-terminal part of the original polyprotein by the action of 3C (Figure 2). An analogous situation exists in the comoviruses. These viruses produce two separately encapsidated RNA molecules, one of which (the M-RNA) codes for the structural proteins. The B-RNA contains the information for the protease, a 24kDa protein which cleaves itself from the N-terminal end of the polyprotein. This protein is also responsible for the separation of the capsid protein from the other polyprotein, but for this function it requires the participation of another B-RNA gene product, the 32kDa protein[6]. By adopting this strategy, both these viruses utilize the same active site in two distinct proteases which separately control the production of capsid proteins and replication enzymes.

There is another level of control over polyprotein cleavage which is exerted at the level of primary structure. All of the sites cleaved by 3C, on its own or in combination with 3D, are glutamine–glycine bonds, but only nine of thirteen such bonds within the polyprotein are cleaved. 2A is even more selective, cleaving at only two of the ten tyrosine–glycine bonds present. Studies using synthetic peptides have suggested that the recognition sequence for the picornavirus 3C protease encompasses a hepta-peptide sequence stretching from five residues to the N-terminal side of the cleaved bond (the P_5 position in the nomenclature of Berger & Schechter[7]) through two residues on the other side $(P_{2'})$[8]. As these sequences are by no means identical at all the cleavage sites, it seems reasonable to assume that the sequence (and/or the secondary structure it adopts) will have an influence on the rate at which the bond will be cleaved. This conclusion is supported by studies that have shown a correlation between the *in vitro* cleavage rate of synthetic peptides corresponding to known cleavage sites to the processing order of these sites *in vivo*. Thus differential rates of proteolysis influenced by the amino acid sequence surrounding the cleavage sites represent an additional means by which viral gene expression can be regulated.

Polyprotein cleavage in other viral families

Recent work on proteases from the Togavirus family, which have RNA genomes within an icosahedral capsid surrounded by an envelope or cloak (toga), has highlighted some interesting refinements of polyprotein processing[9]. Sindbis virus of the genus Alphavirus has a positive-strand RNA genome which acts directly as the message for p270, a polyprotein encompassing the non-structural proteins nsP1–nsP4. The genomic RNA is also transcribed into a full-length negative strand which acts as the

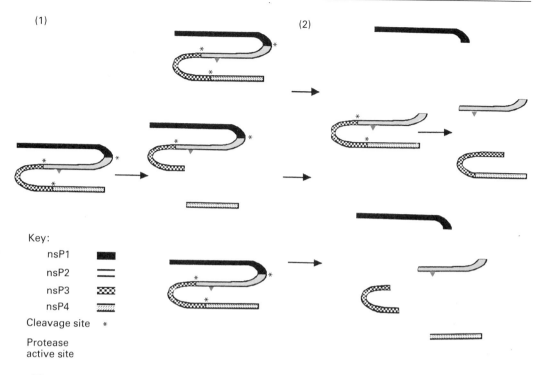

Figure 3. Protease-mediated control of nsP4 expression in Sindbis virus
(1) At low concentration of polyprotein early in infection intramolecular (*cis*) cleavage occurs, releasing nsP4. (2) Later in infection intermolecular (*trans*) cleavages release nsP1 allowing protease access to the nsP2/nsP3 boundary, creating p34 and thus reducing the amount of nsP4 produced. nsP4 is also metabolized by the cellular ubiquitin pathway.

template for a subgenomic mRNA subsequently translated as a 120kDa polyprotein comprising the structural proteins. Synthesis of the negative-sense RNA occurs only in the first few hours after infection and regulation of its synthesis is achieved through protease-mediated control of the replication complex. Of the non-structural proteins arising from p270, nsP2 contains the protease activity while both nsP4 and P34 are components of the replication complex. When nsP4 is present in the complex then both negative and positive strand synthesis occurs, but when P34 displaces nsP4, then only positive strand synthesis occurs.

The substrate specificity of the protease is governed by what it is attached to in the polyprotein. Cleavage at the nsP3/nsP4 boundary requires that nsP2 is still attached to nsP3 while the nsP2/nsP3 junction could not be cleaved if nsP1 was part of the polyprotein. As a consequence of this, early in infection the major products were nsP4 and the P123 polyprotein while later this balance changed to nsP1, nsP2 and a P34 protein (Figure 3). Sindbis virus nsP4 also contains the sequence motifs promoting degradation through the cellular ubiquitin pathway, further decreasing its stability and thereby controlling the synthesis of negative-strand RNA. The 2A-mediated cleavage in the 3D (polymerase) domain of the poliovirus polyprotein may achieve a similar purpose, although mutant viruses where the cleavage site has been altered to a non-cleavable form display a normal phenotype.

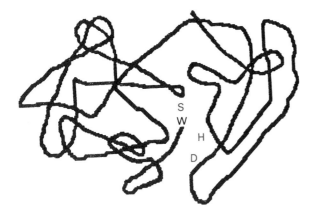

Figure 4. Structure of the Sindbis virus capsid protein
The capsid protein is an autoprotease which releases itself from the poly-protein, creating its own inhibitor in the process. This α-carbon plot shows the two-domain fold of trypsin-like serine proteases. The C-terminal tryptophan (W) formed during autoproteolysis re-mains in the active site of the enzyme close to the serine (S) nucleophile. His-tidine (H) and aspartic acid (D) are the other members of the catalytic triad. Dia-gram adapted from Choi *et al.*[10]

The fate of the other Sindbis polyprotein containing the structural proteins is also interesting. p130 contains the capsid protein at the N-terminus proximal to the three envelope spike proteins. The capsid protein is a serine protease which cleaves itself from the polyprotein while the spike proteins are later separated by cellular proteases. X-ray crystallography studies of the capsid protein by Choi *et al.*[10] reveal that it has the characteristic secondary and tertiary structure of trypsin-like serine proteases. These studies also suggest why it can only perform the one cleavage, that which removes it from the polyprotein, and thereafter can no longer act as a protease. After cleavage from the polyprotein its new C-terminal residue (tryptophan) is left blocking the active site (Figure 4), and dissociation from the rest of the polyprotein leaves one of the other residues involved in the catalytic mechanism (Asp-163) in an environment less conducive to its participation in the interactions with His-141 and Ser-215 that lead to catalysis. The mechanism of action of serine proteases involves the formation of a transient covalent bond between the nucleophilic serine of the protease and the residue which will ultimately form the new C-terminus of the cleaved protein. In the crystal structure of the mature capsid protein, the C-terminal tryptophan is close enough to the serine to suggest that this "transient" bond has become permanent.

Virion maturation

Proteolytic maturation occurs mainly in icosahedral or enveloped viruses. The ma-turation cleavages in the protein capsids of non-enveloped (or naked) virions generally take place after the capsid has been assembled and in most instances after the nucleic acid has been packaged. These cleavages are carried out by virus-coded proteases and are often autocatalytic, the capsid protein doubling as its own maturation protease. In the case of enveloped viruses, maturation involves cleavages of envelope (glyco)pro-teins which are usually carried out by host proteases. In both instances proteolytic maturation is essential for the development of infectious virus.

In the case of the T4 phage the situation is slightly different in that proteolysis is necessary to prepare the capsid (or head) for the packaging of DNA[11]. T4 phage assembly is initiated by the self-association of a so-called scaffolding protein into a core around which the capsid is built. The major capsid protein, p23, then assembles around the core to form a rounded prohead I. At this point the p23 subunits are cleaved by a protease (p21) near to their N-termini and the final icosahedral-like head forms. The scaffolding proteins within the head are then digested by proteases to small peptides, leaving room for the DNA within the capsid. The cleavage of the p23 capsid protein not only leads to changes in the head shape, but also reveals binding sites on the capsid that may aid the entry of the DNA while the peptide released from p23 is involved in packaging the DNA within the capsid.

In poliovirus the final proteolytic event involves the cleavage of VP0 into VP4 and VP2. This occurs simultaneously with the entry of the genome RNA into the capsid and the consequent rearrangement of the capsid may serve to lock the RNA into the virion. Certainly the loss of VP4 during uncoating of the viron after entry into the target cell is accompanied by the release of RNA from the capsid. The protease responsible for the VP0 cleavage has not been identified but it is thought that this is an autocatalytic process.

Adenovirus is a non-enveloped DNA virus that has an absolute requirement for protease activity despite the fact that it does not synthesize a polyprotein. In 1976 Weber[12] reported that a temperature-sensitive mutant of adenovirus (Ad2ts1) which lacked protease activity at the restrictive temperature, formed virus-like particles but these were not infectious, indicating that proteolysis is necessary for maturation of the adenovirus virion. The mutation in Ad2ts1 has been mapped to a 23kDa protein which is synthesized late in infection and this is assumed to be the protease[13]. The 23kDa protein is packaged within the virion and is known to cleave six proteins, some of which are involved in the capsid architecture. While proteolysis is essential for the development of infectivity, it is not clear whether in its absence virions assemble incorrectly, or whether the function of the protease is to destabilize the virion to allow uncoating and release of the nucleoprotein core following uptake into the cell. Adenovirus enters the cell by receptor-mediated endocytosis where it will be exposed to the relatively low pH of the endosome where the major capsid proteins can undergo a conformational change which may be enough to aid the disruption of a capsid weakened by limited proteolysis at key points in its structure.

Proteolysis also plays a role in the maturation of enveloped viruses. Proteins destined for the envelope are synthesized on membrane-bound ribosomes and include at their N-terminus a hydrophobic signal or leader peptide which is cleaved by a host "signalase" once it has initiated binding of the ribosome to the membrane. In general terms the viral envelope proteins, which are frequently glycosylated and often also acylated with palmitic or myristic acid, migrate to their target membrane where they replace the host cell proteins. A viral matrix protein then attaches to the inner surface of the membrane and interacts with the viral proteins, aggregating thereafter and acting as the focus for the attachment of the nucleocapsid. The viral particle is then released by a budding process from the membrane.

In the case of influenza virus the surface glycoprotein haemagglutinin is synthesized as a single chain which must be cleaved to give a disulphide-bonded dimer before the virion can become infectious[14]. The haemagglutinin binds the virus particles to specific host cell receptors and the particles are then endocytosed to "late" endosomes where fusion between the cell membrane and the viral envelope results in the passage of the viral genome into the cytoplasm. In paramyxoviruses, which include measles and mumps viruses, cell attachment and fusion are the province of separate envelope glycoproteins. The haemagglutinin/neuraminidase (HN) protein binds to sialic acid-containing receptors while fusion between the viral envelope and the plasma membrane is controlled by the F (fusion) protein. Proteolytic cleavage of the F protein into F1 and F2 subunits is essential for fusion and therefore infectivity.

Cleavages of the membrane glycoproteins of this type are generally carried out by host-cell proteases, although in some instances bacterial proteases can be responsible with serious consequences for the host. Severe, and often fatal, viral pneumonia can result from simultaneous infection of an individual with certain strains of *Staphylococcus* and with influenza virus[15]. The staphylococcal protease will cleave the viral haemagglutinin, greatly increasing the infectivity of the virus. Further evidence that infectivity is governed by the specificity of host-cell proteases comes from studies on influenza virus derived from mammalian cells which was adapted to grow in chicken embryo cells. The haemagglutinin of the adapted virus was found to contain an insertion of basic amino acids in the cleavage site which then resembled that found in the highly pathogenic avian influenza virus. This haemagglutinin was cleaved by the chicken embryo cell proteases and the adapted mammalian virus caused lethal infection in chickens.

Inactivation of host systems

In some cases virus-coded proteases have a role in facilitating virus processing by inhibition of host cell systems and can also be involved in the processes by which cells are killed during lytic infections. Cell death can result from virus-induced cell–cell fusion to produce syncitia. Syncitia formation occurs in retroviral infections and also in parainfluenza infections where the host protease cleavage of the F protein has a central role.

One of the first events following picornavirus infection of mammalian cells is the shutdown of host cell protein synthesis. This is caused by inhibition of RNA transcription of all three polymerase systems as well as a more direct inhibition of protein symthesis in which both the 2A and 3C proteases are implicated. Eukaryotic mRNAs are capped at their 5' end by methylguanate, and the capped mRNA is recognized by the p25 cap binding protein, a subunit of eukaryotic initiation factor 4F (eiF-4F) complex. eiF-4F then binds the capped RNA to the 40 S ribosomal subunit. Another subunit of eiF-4F, p220, is cleaved following picornavirus infection, inhibiting the action of the complex. The 2A protease participates in this cleavage, although indirectly, in a reaction involving another cellular translation factor, eiF-3[16]. Poliovirus translation proceeds via a cap-independent mechanism and therefore does not require an active eiF-4F complex. Recent work[17] has suggested that 2A also affects host cell protein synthesis by causing a reduction in RNA polymerase II transcription. The picornavirus 3C protease has a role in the inhibition of host cell processes by cleaving the active form of transcription factor IIIC to an inactive form, thereby inhibiting RNA polymerase III transcription[18].

There are various reports that HIV protease will cleave, *in vitro*, cellular proteins such as calmodulin and intermediate filament proteins such as actinin, spectrin, tropomyosin and actin. While it has still to be shown that these events occur *in vivo*, cleavage of cytoskeletal proteins could have a role in the release of virus from the cell.

NATURE OF VIRUS-CODED PROTEASES

All of the proteases so far characterized fall into one of four classes, which take their names from one of the major determinants of catalysis[19]. To date, virus-coded proteases have been found amongst the serine, cysteine and aspartyl proteases, but as yet no viral enzyme of the metalloprotease class has been identified.

Serine proteases, such as the pancreatic enzymes trypsin and chymotrypsin, have a catalytic mechanism dependent on a serine residue which is a potent nucleophile as a result of the combined effects of a histidine and aspartic acid residue brought into close proximity by the characteristic secondary and tertiary structures of the protein. During the cleavage the serine attacks the carboxyl function of the peptide bond, becoming transiently acylated in the process before this unstable intermediate is attacked by water. Serine proteases of the trypsin-like superfamily can be recognised by a conserved motif (GDSGG) around the active site serine and also have a distinctive pattern of secondary structure consisting of 12 strands of β-sheet and a single helix at the C-terminus. Some bacterial serine proteases also fit this general pattern, although they are somewhat smaller proteins, but there is a structurally distinct subset of serine proteases represented by another bacterial enzyme, subtilisin. This has evolved an apparently identical catalytic triad set into a different protein framework — an example of convergent evolution. The sequences surrounding the active site residues are different and the folding pattern of the protein does not resemble that of the trypsin family.

The cysteine proteases, of which the enzyme papain is the archetype, have a reaction mechanism which is thought to involve a cysteine and histidine residue forming a thiolate imidazolium ion pair. This creates a reactive thiolate anion which will attack the carboxyl function of the peptide bond, becoming transiently acylated in an analagous manner to the serine of serine proteases.

Aspartyl proteases, formerly referred to as acid proteases, utilize the interactions of two aspartic acid residues in their catalytic mechanism. One of the aspartate residues is maintained in the undissociated form which contributes to the generally acidic pH optimum of these enzymes. Each aspartate is located within an Asp-Thr-Gly (DTG) sequence located in separate but very similar domains of the molecule. Occasionally serine replaces threonine.

The most extensively studied viral proteases are the 2A and 3C proteases of picornaviruses, notably the polio- and rhinovirus enzymes and the proteases of retroviruses, particularly that of HIV. The HIV protease is a member of the aspartyl family but is unusual in that it is a small protein of 99 amino acids with only one DTG sequence. The catalytic mechanism demands two aspartate residues and therefore the protease must be active as a dimer. This raises interesting questions as to how it contrives the initial cleavages necessary to release itself from the polyprotein, but nevertheless it does, and the requirement for dimerization represents an important aspect of control. The major proteolytic events carried out by the retroviral proteases take place after the polyproteins are enclosed in the viral envelope and after the immature virion has budded from the infected cell. Before budding the polyproteins attach themselves to the inner surface of the membrane through a myristoyl group at their N-terminus and the consequent proximity of polyprotein molecules will increase the possibility of dimerization. In this method of viral assembly, premature cleavage of the polyprotein would be detrimental, as it would allow the separated proteins to disseminate within the cell instead of focussing at the appropriate location via the myristoyl anchor. This was neatly demonstrated by Krausslich[20] who engineered a second protease monomer into the polyprotein allowing the formation of active protease within a single molecule. In an *in vitro* translation system the polyprotein containing the dimer was rapidly processed in a concentration-independent manner. This contrasts with the wild type where the polyprotein is poorly processed in such translation systems. Following transfection of the constructs no viral particles were observed, consistent with premature processing of the polyprotein preventing assembly, and in addition there was a marked cytotoxicity lending further weight to the suggestion that the protease cleaves cellular proteins.

The crystal structure of the HIV protease revealed another significant difference from the pepsin-like aspartyl proteases. There is, in each monomer, a region of less-well-defined structure apparently covering the access to the active site cleft. These so-called flap regions are better defined in crystals formed in the presence of some inhibitors and there are strong indications that they have a role in the binding of inhibitor, and presumably substrate. Two strands of the flap region, the inhibitor and a stretch of sequence immediately following the active site DTG, form a four-stranded β-sheet structure on either side of where the susceptible bond would be[21] (Figure 5). This is consistent with the observation that the sites known to be cleaved by retroviral proteases do not fit any readily recognized pattern of primary structure, but the recognition sequence extends over at least seven residues. In this case secondary structure is at least as important as primary structure in defining sites that are cleaved by this protease.

Inhibition studies and site-directed mutagenesis suggest that the picornavirus proteases are of the cysteine class. However, the relative positions of the putative active site histidine and cysteine residues are different from that found in the papain family. Bazan & Fletterick[22] and Gorbalenya *et al.*[23] both showed that the predicted secondary structure of 3C resembled the characteristic pattern found in serine proteases such as trypsin. In a similar way the 2A protease was found to resemble the smaller, bacterial trypsin-like proteases. The current hypothesis is that these proteases represent the founder members of a novel subclass of cysteine protease that has a reaction mechanism similar to that of trypsin but with a cysteine instead of a serine as the nucleophile in the catalytic triad. The alignments produced by Bazan & Fletterick and Gorbalenya *et al.* agree on the assignment of Cys-147 and His-40 as two of the catalytic triad but they differ on the identity of the participating acidic residue. Bazan & Fletterick favour Asp-85 while Gorbalenya *et al.* suggest Glu-71. Substitution, by site-directed mutagenesis, of His-40, Glu-71 and Cys-147 with tyrosine, glutamine and serine respectively produced an inactive enzyme, while replacement of the highly conserved Asp-85 with asparagine produced a protease capable of cleaving protein and polypeptide substrates. The situation is still not clear, however, as the aspartate-to-asparagine mutant, although active in *trans* cleavages (those where the cleaved bond is in another protein molecule) was unable to cleave itself from the polyprotein[24].

Work from our laboratory based on inhibitor studies has suggested that the adenovirus protease is also of the cysteine class[25], although it does not fit into the category

Figure 5. The AIDS virus protease
An α-carbon plot of the protease from
HIV-1, drawn from Brookhaven Data
Bank co-ordinates of protease crystallized in the presence of a peptide-
derived inhibitor (shown schematically
as the black arrows). The flap regions
(open arrows to the right) move at least
7Å on binding the inhibitor to form two
four-stranded β-sheets with the inhibitor
and a strand of the protease immediately following the active site aspartic
acid residues (*). The major dimerization
contacts are at the N- and C-termini, the
active site region and the flaps.

of trypsin-like cysteine proteases. The protease gene from adenovirus has been sequenced in seven human and two bovine serotypes and when the derived protein
sequences are aligned, there is only one conserved histidine and three conserved
cysteines. None of these are in recognizable protease sequence motifs and the conserved histidine is closer to the N-terminus than any of the conserved cysteines —
the opposite order to that found in papain. Searches of the 23kDa sequence against
various protein sequence data bases fails to reveal any significant sequence similarity
with any known protease, providing further evidence for the unusual nature of this
enzyme. The cleavage specificity determined by synthetic peptide studies[26] is influenced mainly by the residues in the P_4 and P_2 positions. The residue at P_4 must
be either methionine, leucine or isoleucine while the P_2 position must be occupied
by glycine. Interestingly this specificity is similar to that determined for the as-yet-

unidentified protease from another DNA virus, vaccinia[27].

Cysteine proteases regarded as papain-like on the basis of sequence alignments have been identified in a number of viruses (Table 1). Likewise, serine proteases have been identified, although these are less common, being found most notably in the Togavirus family. The Sindbis virus capsid autoprotease is a serine protease, although mutagenesis of the active site serine to a cysteine or threonine still permits cleavage to occur suggesting that these, too, are unusual proteases.

PROSPECTS FOR PROTEASE INHIBITORS AS ANTI-VIRAL AGENTS

The essential role of viral proteases in the development of infectivity allied to their often unusual nature as exemplified by the retroviral, picornaviral and adenoviral proteases make them attractive targets for antiviral therapy. The pressing need for an effective treatment for AIDS has led to a particularly intensive study of the HIV protease and a concerted search for effective inhibitors. One approach followed by several groups is to develop substrate peptides into non-cleavable substrate analogues, or, more effectively, analogues of the transition state between substrate and products. This approach has resulted in inhibitors that are effective at the nanomolar level and that show selectivity for viral aspartyl proteases, failing to inhibit significantly the mammalian aspartyl proteases such as pepsin at micromolar concentrations. While problems of delivery and targeting remain to be addressed, they have been shown to be effective in inhibiting viral replication in cell culture and have a low cytotoxicity[28].

The existence of a crystal structure of the HIV protease has led to another approach to the development of effective therapeutic inhibitors[29]. Computer analysis was used to develop a topographical map of the active site of the protease, and this map was used to scan a database of molecular structures in the hope of finding a molecule with an appropriate complementary shape which would lock onto the enzyme. This approach was successful in that the compound Haloperidol was found to have steric complementarity and indeed did inhibit the protease with a K_i of 100 μM. Unfortunately, Haloperidol is highly toxic at concentrations useful for protease inhibition. However, this work does suggest an approach that is useful in identifying structures worthy of further investigation.

In order to be active HIV protease must form dimers. The main contact areas are at the terminal regions of the molecule which form an interdigitating β-sheet, at the active site and in the region of the flaps (Figure 5). Inhibition of this dimerization presents an alternative target for protease-directed therapeutic agents.

SUMMARY

● Proteases were amongst the first enzymes to be isolated and crystallized, and the discovery of their existence in viruses and the realization of the vital role they play has given a new lease of life to one of the oldest topics in biochemistry.

● Already we have seen the discovery of new variations on well-studied reaction mechanisms and there is the promise of others to come that may be totally novel.

● When this is allied to the prospect of developing the knowledge which is beginning to accrue into a much needed antiviral therapy, it is clear that viral proteases and their key role in viral replication will be an increasing focus of attention for some time to come.

REFERENCES

1. Krausslich, H.G. & Wimmer, E. (1988) Viral proteinases. *Annu. Rev. Biochem.* **57**, 701–754
2. Riechmann, J.L., Lain, S. & Garcia, J.A. (1992) Highlights and prospects of potyvirus molecular biology. *J. Gen. Virol.* **73**, 1–16
3. Wong-Staal, F. & Vaishnav, Y.N. (1991) The biochemistry of AIDS. *Annu. Rev. Biochem.* **60**, 577–630
4. Kay, J. & Dunn, B.M. (1990) Viral proteinases: weakness in strength. *Biochim. Biophys. Acta* **1048**, 1–18
5. Hellen, C.U.T., Krausslich, H.G. & Wimmer, E. (1989) Proteolytic processing of polyproteins in the replication of RNA viruses. *Biochemistry* **28**, 9881–9890
6. Vos, P., Verver, J., Jaegle, M., Wellink, J., Van Kammen, A. & Goldbach, R. (1988) Two viral proteins involved in the proteolytic processing of the cowpea mosaic virus polyproteins. *Nucleic Acids Res.* **16**, 1967–1985
7. Berger, A. & Schechter, I. (1970) Mapping the active site of papain with the aid of peptide substrates and inhibitors. *Philos. Trans. R. Soc. London Ser. B* **257**, 249–264
8. Long, A.C., Orr, D.C., Cameron, J.M., Dunn, B.M. & Kay, J. (1989) A concensus sequence for substrate hydrolysis by rhinovirus 3C proteinase. *FEBS Lett.* **258**, 109–112
9. Strauss, J.H. & Strauss, E.G. (1990) Alphavirus proteinases. *Semin. Virol.* **1**, 347–356
10. Choi, H.K., Tong, L., Minor, W., Dumas, P., Boege, U., Rossmann, M.G. & Wengler, G. (1991) Structure of Sindbis virus core protein reveals a chymotrypsin like serine proteinase and the organisation of the virion. *Nature (London)* **354**, 37–43
11. Black, L.W. & Showe, M.K. (1983) Morphogenesis of the T4 head, in *Bacteriophage T4* (Mathews, C.K., Kutter, E.M., Mosig, G. & Berget, P., eds.), pp. 219–245, American Society of Microbiology, Washington DC
12. Weber, J. (1976) Genetic analysis of adenovirus type 2. III Temperature sensitivity of processing of viral proteins. *J. Virol.* **17**, 462–471
13. Yeh-Kai, L., Akusjarvi, G., Alestrom, P., Pettersson, U., Tremblay, M. & Weber, J. (1983) Genetic identification of an endoproteinase encoded by the adenovirus genome. *J. Mol. Biol.* **167**, 217–222
14. Wiley, D.C. & Skehel, J.J. (1987) The structure and function of the hemagglutinin membrane glycoprotein of influenza virus. *Annu. Rev. Biochem.* **56**, 365–394
15. Tashiro, M., Ciborowski, P., Klenk, H.D., Pulverer, G. & Rott, R. (1987) Role of staphylococcal protease in the development of influenza pneumonia. *Nature (London)* **325**, 536–537
16. Wyckoff, E.E., Hershey, J.W. & Ehrenfeld, E. (1990) Eukaryotic initiation factor 3 is required for poliovirus 2A protease induced cleavage of the p220 component of eukaryotic initiation factor 4F. *Proc. Natl. Acad. Sci. U.S.A.* **87**, 9529–9533
17. Davies, M.V., Pelletier, J., Meerovitch, K., Sonenberg, N. & Kaufman, R.J. (1991) The effect of poliovirus proteinase 2Apro expression on cellular metabolism: inhibition of DNA replication, RNA polymerase II transcription and translation. *J. Biol. Chem.* **266**, 14714–14720
18. Clark, M.E., Hammerle, T., Wimmer, E. & Dasgupta, A. (1991) Poliovirus proteinase 3C converts an active form of transcription factor IIIC to an inactive form: a mechanism for inhibition of host cell polymerase III transcription by poliovirus. *EMBO J.* **10**, 2941–2947

19. Barrett, A.J. (1977) Introduction to the history and classification of tissue proteinases, in *Proteinases in Mammalian Cells and Tissues* (Barrett, A.J., ed.), pp. 1–55, North-Holland, Amsterdam

20. Krausslich, H.G. (1991) Human immunodeficiency virus proteinase dimer as component of the viral polyprotein prevents particle assembly and viral infectivity. *Proc. Natl. Acad. Sci. U.S.A.* **88**, 3213–3217

21. Gustchina, A. & Weber, I.T. (1990) Comparison of inhibitor binding in HIV-1 protease and in non viral aspartic proteases: the role of the flap. *FEBS Lett.* **269**, 269–272

22. Bazan, J.F. & Fletterick, R.J. (1988) Viral cysteine proteases are homologous to the trypsin-like family of serine proteases: structural and functional implications. *Proc. Natl. Acad. Sci. U.S.A.* **85**, 7872–7876

23. Gorbalenya, A.E., Donchenko, A.P., Blinov, V.M. & Koonin, E.V. (1989) Cysteine proteases of positive strand RNA viruses and chymotrypsin-like serine proteases. A distinct superfamily with a common structural fold. *FEBS Lett.* **243**, 103–114

24. Hellen, C.U.T, Facke, M., Krausslich, H.G., Lee, C.K. & Wimmer, E. (1991) Characterisation of poliovirus 2A proteinase by mutational analysis: residues required for autocatalytic activity are essential for induction of cleavage of eukaryotic initiation factor 4F polypeptide p220. *J. Virol.* **65**, 4226–4231

25. Webster, A., Russell, W.C. & Kemp, G.D. (1989) Characterisation of the adenovirus proteinase: development and use of a specific peptide assay. *J. Gen. Virol.* **70**, 3215–3223

26. Webster, A., Russell, S., Talbot, P., Russell, W.C. & Kemp, G.D. (1989) Characterisation of the adenovirus proteinase: substrate specificity. *J. Gen. Virol.* **70**, 3225–3234

27. VanSlyke, J.K., Whitehead, S.S., Wilson, E.M. & Hruby, D.E. (1991) The multistep proteolytic maturation pathway utilised by vaccinia virus P4a protein: a degenerate conserved cleavage motif within core proteins. *Virology* **183**, 467–478

28. Roberts, N.A., Martin, J.A., Kinchington, D., Broadhurst, A.V., Craig, J.C., Duncan, I.B., Galpin, S.A., Handa, B.K., Kay, J., Krohn, A., Lambert, R.W., Merrett, J.H., Mills, J.S., Parkes, K.E., Redshaw, S., Ritchie, A.J., Taylor, D.L., Thomas, G.J. & Machin, P.J. (1990) Rational design of peptide based HIV proteinase inhibitors. *Science* **248**, 358–361

29. Desjarlais, R.L., Seibel, G.L., Kuntz, I.D., Furth, P.S., Alvarez, J.C., OrtizdeMontellano, P.R., DeCamp, D.L., Babe, L.M. & Craik, C.S. (1990) Structure based design of nonpeptide inhibitors specific for the human immunodeficiency virus 1 protease. *Proc. Natl. Acad. Sci. U.S.A.* **87**, 6644–6648

2

The biochemistry of blood clotting : the digestion of a liquid to form a solid

Michael F. Scully

Thrombosis Research Institute, Emmanuel Kaye Building, Manresa Road, Chelsea, London SW3 6LR, U.K.

INTRODUCTION

The fluidity of blood is remarkable. Blood, half the volume of which is red cells, flows through the vasculature at rates of between 1mm/s in the capillary vessels (which are of a diameter similar to the red cells) to 300 mm/s in the main arteries and at wall shear rates of up to 3000 s^{-1}. Considering such a property it is perhaps even more remarkable that a blood sample, taken by venepuncture and left to stand, will change within a few minutes to a tough, jelly-like coagulum. This transition is due, of course, to blood coagulation, the mechanism by which the integrity of this high pressure, closed loop, circulatory system is maintained after blood vessel injury. When a blood vessel is cut or damaged the injury initiates a series of events which lead to the prompt formation of a blood clot to curtail the haemorrhage (haemostasis). In the first stage of haemostasis a constriction of the vessel occurs together with the rapid formation of a temporary haemostatic plug consisting of platelet cells (Figure 1). Subsequently the plug is transformed into a stabilized clot, by the conversion of a soluble plasma protein into an insoluble one. The latter spontaneously polymerizes into long strands which anneal into a vast network to coagulate the blood. *In vivo* the activity of the clotting mechanism responsible for this conversion is controlled and localized to the immediate area of damage, thus preventing clots from developing in uninjured vessels and maintaining the circulating blood in a fluid state.

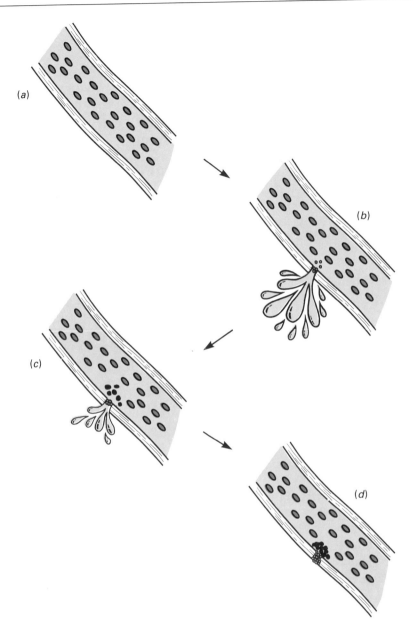

Figure 1. Course of events during the termination of bleeding from a vessel
When a small blood vessel is punctured (a) the escape of blood is slowed within seconds by
the rapid formation of a plug consisting of platelet cells (grey) (b). Immediately proteolytic clotting
factors are activated which generate the insoluble protein, fibrin. Fibrin (black) is formed only
within the locality of the bleeding site and polymerizes spontaneously into a vast network
capable of trapping blood cells and serum (c). This clotted mass contracts and binds together
at the site thereby completely halting the haemorrhage (d). In normal healthy individuals, when
bleeding is started by a controlled incision, the rate of blood flow reaches a peak after 1 minute,
reduces rapidly and is halted completely after 7 minutes or so. Evidence of rapid activation of
platelets and formation of fibrin is found in samples taken within 10 seconds of the incision[27].

Our knowledge of the mechanisms by which blood clotting occurs and is controlled has increased greatly over the last decade, principally through publication of the protein structure of the known coagulation factors, investigation of the enzymology of blood coagulation and through characterization of control systems[1-4].

In spite of this increase in knowledge, thrombosis ("haemostasis in the wrong place") still accounts for a very high proportion of deaths in the Western world (25% of all deaths in the U.S.A. each year) and there is continuing need to develop further our understanding of this crucial aspect of mammalian physiology.

Herein is given an overview of current understanding of the clotting mechanism but without consideration of the role and properties of blood platelets which is dealt with elsewhere[5]. Perhaps because of the intimate association between the blood and organs of the body, the proteins of haemostasis have also been found to be involved in aspects of normal tissue physiology and in the progress of some non-thrombotic disease (for example cancer and inflammation)[6], properties which are, also, not considered in this article.

ESSENTIAL FEATURES OF BLOOD CLOTTING

Blood clotting is an ever-ready system for plugging holes which is activated when needed. When bleeding occurs a molecular switch is thrown at the leakage point, generating the formation of a fast-acting, filling agent (fibrin) which is pumped into the site until the bleeding stops (Figure 1). Then the switch is turned off and the production of the filler abruptly ceases. The proteins responsible for this mechanism, the blood clotting factors, are present in the plasma (the soluble part of blood) in inactive forms. A stimulus is needed before clotting can begin to occur and this can be the contact of blood with a negatively charged surface (termed, historically, the intrinsic pathway of coagulation) or contact with tissue factor, a glycoprotein found in the membranes of certain cells (the extrinsic pathway). These stimuli initiate a limited proteolysis of each of the blood clotting factors to generate the active forms (hence the subtitle to this Essay: "the digestion of a liquid to form a solid").

In order to amplify the initial stimulus the clotting factors are activated in a sequence more or less in accordance with their concentration in blood. Because of this concentration gradient one molecule of the first proteolytic factor generates ten molecules of the next, which generates a hundred of the next; a crescendo of proteolytic activity. This sequence appeared, to those who first described it, to resemble a cascade and the inter-relationships between the various clotting factors, as shown in Figure 2, have therefore become to be known as the coagulation cascade. At some of the proteolytic steps the rate of conversion is dependent on the presence of a cell surface and calcium ions (for which reason a blood sample can be anticoagulated by the simple addition of a calcium chelator such as sodium citrate or EDTA) and is potentiated by binding to activated, non-enzymic cofactors, factor Va and factor VIIIa (activated forms of each factor are noted by the addition of the suffix "a", hence V to Va). Clot formation occurs when a soluble blood protein, fibrinogen, is converted by thrombin (factor IIa) to an insoluble one, fibrin monomer, which assembles spontaneously into a polymer.

The question then arises as to why clotting once started does not propagate through-

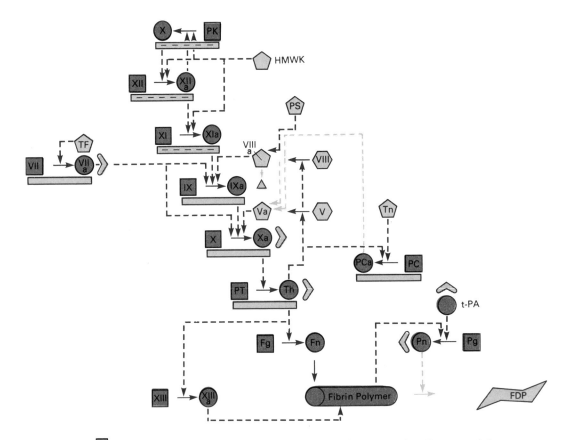

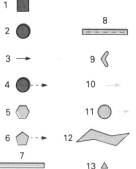

Figure 2. The coagulation cascade: diagram of the reactions occurring during blood coagulation

Abbreviations: coagulation factors are named according to their Roman numerical title, for example X, coagulation factor X; Xa, activated coagulation factor X; TF, tissue factor; PK, prekallikrein; K, kallikrein; PC, PCa, protein C; PS, protein S; PT, prothrombin; Th, thrombin; Fg, fibrinogen; Fn, fibrin; FDP, fibrin digestion products; Pg, plasminogen; Pn, plasmin; Tn, thrombomodulin; t-PA, tissue type plasminogen activator; HMWK, high-molecular-weight kininogen. Key: 1, inactive and 2, active forms of blood clotting factor; 3, proteolytic activation of clotting factor; 4, proteolytic factor responsible for activation of clotting factor; 5, inactive and 6, active non-enzymic cofactors of proteolytic reactions; 7, membrane surface phospholipid acting as cofactor in the presence of Ca^{2+} by localizing activation step; 8, negatively charged surface responsible for contact activation pathway of coagulation; 9, proteinase inhibitors of clotting factors: antithrombin III inhibits thrombin and factor Xa, tissue factor pathway inhibitor inhibits tissue factor/factor VIIa complex; plasminogen activator 1 inhibits t-PA, and α_2-antiplasmin inhibits plasmin; 10, proteolytic degradation of clotting factor to stop further coagulation; 11, proteolytic factor responsible for degradation of clotting factors; 12, plasmin digestion products of fibrin; 13, protein Ca digestion products of factor Va and VIIIa.

out the entire blood system with fatal consequences. The answer to this question still awaits a final solution but it is clear that in the coagulation cascade there are a number of mechanisms which act as positive and negative feedback loops. The activation of the coagulation proteins only occurs in the presence of an initial stimulus and is exceedingly slow in the absence of the necessary surfaces and cofactors, properties which act to limit and localize the clotting. Although coagulation will occur rapidly at and about the site of injury the momentum of the reactions will diminish as the distance from the stimulus increases. Furthermore, the clotting enzymes are liable to inhibition by specific inhibitors and the products of clotting activate powerful proteolytic systems which digest fibrin and also the activated cofactors, thereby acting as a negative feedback loop to shut down the entire mechanism.

THE STRUCTURE OF BLOOD CLOTTING FACTORS: MOLECULES DESIGNED FOR A SWITCH SYSTEM

The structure of the clotting factors is the key to full understanding of the mechanism by which clotting occurs and is controlled. Research in this area has given fascinating insight into the evolution of proteins elegantly adapted to the requirement of this complex physiological system, which must always be in a perfect state of balance for the wellbeing of the organism.

A non-descriptive nomenclature has arisen to assign individual blood coagulation proteins. This is largely for historical reasons since they were first characterized as unknown components which corrected the abnormal clotting times of patients with bleeding problems. Most are known by a Roman numeral (e.g. coagulation factor V, shortened to factor V) or by reference to the position of the protein in a chromatographic elution profile (for example, protein S). Structurally, the proteins show considerable evidence of gene duplication, gene modification and exon shuffling during their evolution (Figure 3)[7]. The three-dimensional structures of these homologous proteins are likely to be nearly identical with substitution of amino acid side chains on the protein surface giving each factor its unique properties in regard to substrate recognition and membrane interaction[8].

The catalytic domain of the coagulation proteinases is of the serine proteinase type with structural similarity to that of trypsin/chymotrypsin. The enzymes occur in blood in their zymogen or proenzyme form and are converted to the active form by proteolysis of one or two peptide bonds. The synthesis of the mature form of some of the proteinases is dependent upon vitamin K which is a coenzyme for γ-glutamyl-carboxylase. This enzyme catalyses the carboxylation of between ten and twelve glutamic acid residues in the N-terminus of the molecule (the Gla-domain). Two to three of each of the γ-carboxyglutamyl (Gla) residues bind a single calcium ion, causing a conformational change in the protein which exposes the cell membrane binding site. It is possible that these residues also bind directly to the membranes through calcium ion bridges. It is these effects of calcium which are blocked when chelators are added to blood samples to prevent clotting. The pharmaceutical coumarin drugs which are used for long-term treatment of patients at risk of thrombosis also interfere with the calcium-binding properties of these factors. As analogues of vitamin K they block the γ- carboxylation reaction. Coumarin is also marketed as the rat poison Warfarin,

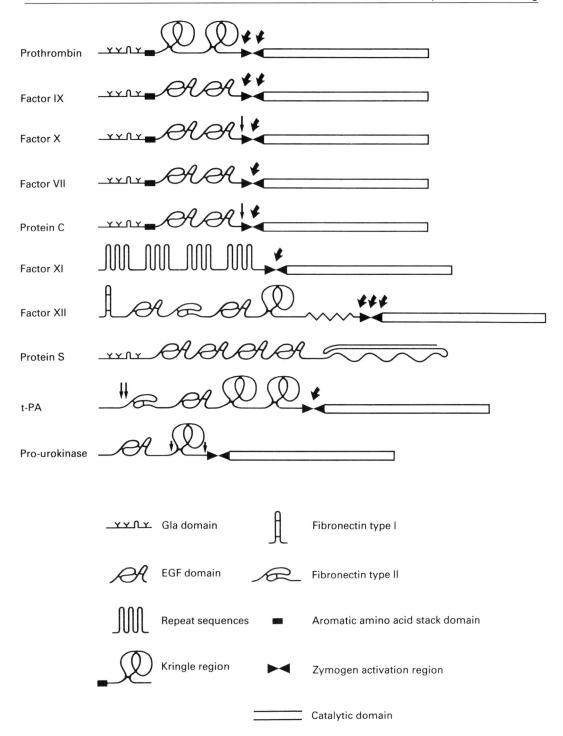

Figure 3. Structural domains of some of the proteins involved in blood coagulation (adapted from[3])

Sites of proteolytic cleavage associated with zymogen activation are indicated by arrows.

which induces fatal intestinal bleeding.

Many of the proteins involved in haemostasis have a "kringle" domain (Figure 3) consisting of 80–85 amino acid residues constrained in a characteristic triple loop by three disulphide bridges (named kringles because of their resemblance to a type of Scandinavian cake). The overall shape of the kringle is discoid, arising from the highly conserved internal amino acids, differences in functional properties being conferred by surface features. Another domain, structurally similar to the epidermal growth factor precursor, is found in a number of the coagulation factors and may be responsible for binding to cell surfaces or to receptor domains on other proteins. Lastly, there are two domains similar to regions in one of the extracellular "sticky" molecules, fibronectin. These domains appear to be involved in binding of the proteins to deposited, insoluble molecules such as fibrin and collagen.

THE FORMATION OF POLYMERIZED FIBRIN: MOLECULAR NETTING

Fibrinogen is the precursor of fibrin, the protein responsible for actually blocking the leaks during blood clotting[9]. Fibrinogen is freely soluble but when treated with thrombin a partial proteolysis occurs with the release of small peptides from the N-terminus. The partial digestion generates fibrin monomer, a molecule which is very insoluble because of its tendency to aggregate spontaneously into large multimolecular aggregates. This tendency arises from the exposure of polymerization zones on the E domain (at the N-terminus) which interact with available polymerization zones on the two D domains [towards the C-terminus of the molecule (Figure 4)]. The interaction permits a half-staggered end-to-end binding of monomers (Figure 4) which extends linearly to form long strands, whose lateral interaction forms the polymer mass. The rapid formation of this molecular netting traps red and white cells, platelets and serum (plasma without fibrinogen) into a clotted mass. The plug is stabilized to proteolysis and the shear forces of the circulating blood by crosslinking of the fibrin network. Isopeptide bonds are formed between glutamic acid and lysine residues [γ-(γ-glutamyl)-lysine] in adjacent fibrin monomers in a reaction catalysed by a calcium-dependent transglutaminase (coagulation factor XIIIa) present in an inactive form in blood and activated by thrombin.

PROPERTIES OF THROMBIN, THE CLOTTING ENZYME: TRYPSIN IN A MUZZLE

Thrombin is the central bioregulatory enzyme in haemostasis, functioning at all levels from the fluid to the cellular level[10]. The synthesis of its precursor, prothrombin, occurs in the liver (like that of all but a few of the coagulation proteins) and requires vitamin K for the post-ribosomal carboxylation of glutamic acid residues in the N-terminus of the zymogen. During the formation of thrombin from prothrombin by factor Xa there are two cleavages. The first, in the catalytic domain of the molecule, generates the active enzyme; the second cleavage releases an activation peptide containing the calcium-binding regions, thus liberating the enzyme from the cell membrane to which prothrombin is attached via the calcium sites during its activation.

As an enzyme, thrombin is a paradigm for all the other coagulation proteinases in that, as compared to trypsin, it exhibits a restricted specificity. Thus thrombin digests

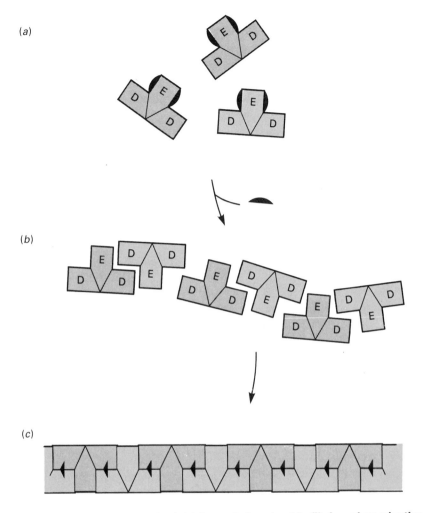

Figure 4. Schematic outline of the initial events involved in fibrin polymerization
Fibrinogen, a soluble plasma protein is liable to proteolysis by thrombin with the release of small peptides (a). This digestion to fibrin monomer exposes polymerization zones on the E domain of the molecule which interact with D domains (the C-terminus) on other fibrin monomer (N-terminus) molecules (b). Half-staggered interactions between these domains form long fibrin strands which are stabilized by crosslinking of D domains between molecules. The covalent crosslinks (♦) are formed by the transglutaminase, factor XIIIa, generated from an inactive precursor by the action of thrombin. The strands interact also laterally and crosswise, generating a vast network trapping red blood cells and serum.

only a limited range of physiological substrates but at high catalytic rate (for the formation of fibrin monomer from fibrinogen $k_{cat.}/K_m$ is 10^7 M^{-1} s^{-1}). Trypsin, of course, can digest proteins to small peptides by cleavage of all available Lys-Xaa or Arg-Xaa peptide bonds. Thrombin has the same catalytic apparatus as trypsin, it cleaves at Arg-Xaa, and yet it selects only four such bonds in fibrinogen, a molecule which contains in total 136 Arg-Xaa bonds. The difference in activity exhibited by the two enzymes is due to the structure of the active site of thrombin which restricts the binding access of potential substrates. The nature of the active site has been reported in recent crystallographic studies which demonstrate a number of specific structures on the surface of the molecule, in particular a narrow active site cleft[11].

The influence of thrombin upon haemostasis is very great since by activating the essential cofactors factor V and factor VIII it acts as a catalyst for its own formation (positive feedback loop), and by activating protein C it engenders a proteolytic activity which digests and destroys factor Va and factor VIIIa, thus turning off the formation of more thrombin (see below). Thrombin is also a potent agonist of platelet activation and has cytokinic and mitogenic properties with a variety of cell types. Recently, a functional thrombin cell receptor has been cloned which exhibits a novel mechanism of activation, involving a proteolytic cleavage[12].

MECHANISMS RESPONSIBLE FOR THE RAPID, LOCAL GENERATION OF THROMBIN: THE MOLECULAR ON-SWITCH

Thrombin is a highly efficient enzyme towards its physiological substrate and at a very low concentration (10 nM, approximately 0.5% of the blood concentration of its precursor, prothrombin) it will clot blood in about 20 seconds. In order that haemostasis is effective without instigating clotting within the vasculature at areas away from the leakage point in the blood vessels (to avoid progression to thrombosis) it is essential that there is a prompt response at the point of haemorrhage by the rapid and localized generation of activated clotting factors from inactive precursors and that this response should be quickly curtailed when bleeding has been halted. What are the properties of the clotting proteins that ensure that this can occur?

Amplification of the initial stimulus: the cascade mechanism

In 1964 the organization of the known clotting factors into the coagulation cascade was proposed independently by MacFarlane and by Davie and Ratnoff (see [1]). A key factor of the proposed cascade was that the initial stimulus, causing activation of factor XII (intrinsic pathway) or factor VII (extrinsic pathway) (Figure 1), would be amplified during the subsequent, sequential activation of the other coagulation factors. In order to act in this manner each clotting factor should, firstly, be specific for activating the next factor in the cascade. This is now well established. For example, factor IXa in the presence of factor VIIIa, cell membrane (or a phospholipid substitute containing the negatively charged phosphatidylserine) and calcium is a potent activator of factor X but the constraint upon its activity is such that it will not readily cleave any other substrate (even small synthetic substrates based upon the peptide sequence cleaved during the activation of factor X). Secondly, the factors should be present in an ascending order of concentration (moving towards the activation of prothrombin) so that each activation will amplify the signal. This again has been confirmed since the concentration gradient from factor VII to prothrombin is now known to be over two orders of magnitude (Table 1).

Localization: the importance of surfaces

Most reactions of the clotting cascade are catalysed by enzyme complexes assembled on a surface together with a non-enzymic cofactor. In the first part of the extrinsic pathway of coagulation (termed the contact phase[13]) there is assembly of factor XII, factor XI and prekallikrein together with the cofactor high-molecular-weight kininogen in close association with surfaces having a net negative charge (for example, glass: the nature of the negatively charged surface responsible under physiological conditions is not clear but may include extravascular tissues rich in collagen and/or sulphatide). Upon binding to the negatively charged surface, factor XII becomes more susceptible to activation and, by an as-yet-unclarified mechanism, factor XIIa is generated. Factor XIIa then activates prekallikrein to kallikrein which can activate factor XII; these reciprocal activations cause an eruptive generation of factor XIIa and subsequently of factor XIa. Factor XIa is able to activate factor IX so that thrombin formation is an ultimate product of the intrinsic pathway. It should be said that the physiological significance of the intrinsic pathway as regards haemostasis is not clear and a number of observations suggest a key involvement of the contact system in the inflammatory and host defence systems. In fact, the patient in whom a deficiency of factor XII was first discovered suffered very little with bleeding problems and died of a heart attack! Consumption and depletion of contact factors is observed in

Table 1. Proteins involved in blood clotting

The approximate number of molecules of each protein per unit volume of blood are expressed relative to that of proteins at the lowest concentration (factor VIII and tissue plasminogen activator), which are set at 1. The actual concentration (M) is given in parentheses.

Protein	Molecular mass (kDa)	Number of molecules per unit volume
Clot makers		
Fibrinogen	340	100000 (10^{-5})
Factor XIII	300	10000 (10^{-6})
Clotting proteinases		
Prothrombin	72	10000 (10^{-6})
Factor X	56	1000 (10^{-7})
Factor IX	56	1000 (10^{-7})
Factor VII	50	100 (10^{-8})
Factor XI	160	100 (10^{-8})
Factor XII	80	1000 (10^{-7})
Prekallikrein	85	1000 (10^{-7})
Cofactors		
Factor V	330	100 (10^{-8})
Factor VIII	330	1 (10^{-10})
High-molecular-weight kininogen	120	1000 (10^{-7})
Inhibitors		
Protein C	62	1000 (10^{-7})
Protein S	80	1000 (10^{-7})
Antithrombin III	65	10000 (10^{-6})
Tissue factor pathway inhibitor	38	100 (10^{-8})
Fibrinolytic system		
Plasminogen	89	20000 (10^{-6})
α_2-Antiplasmin	70	10000 (10^{-6})
Tissue plasminogen activator	65	1 (10^{-10})
Plasminogen activator inhibitor	50	10 (10^{-9})

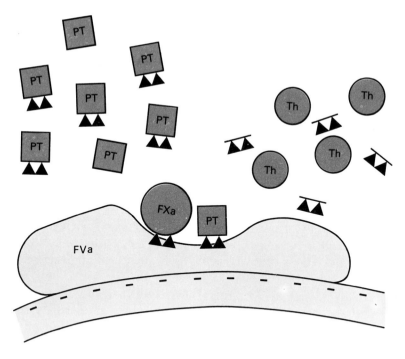

Figure 5. The formation of thrombin at the cell surface
When factor Va binds to a cell surface rich in phosphatidylserine (a negatively charged phospholipid) it forms a high-affinity site for binding factor Xa molecules in complex with Ca^{2+} (bound through γ-carboxyl groups at the N-terminus, ▲). The assembled prothrombinase complex rapidly activates prothrombin, also in complex with calcium ion. At the normal level of prothrombin in blood, the activation of about 0.2% of the blood level of factor X to factor Xa forms enough prothrombinase complex to generate, in 1 second, enough thrombin to clot blood within 10 seconds. The calculated time needed for this concentration of free factor Xa (without these cofactors) to generate the same amount of thrombin would be 3.5 days! Some prothrombin molecules are also shown here without the γ-carboxyl groups (as is the case in blood of patients undergoing warfarin therapy) and are not liable to rapid activation by this complex.

patients suffering from cancer, endotoxaemia, sepsis and allergic reaction. The pain-inducing peptide, bradykinin, is released from high-molecular-weight kininogen by the action of kallikrein and is a potent vasodilator.

Formation of factor IXa constitutes a common junction of the intrinsic and extrinsic pathways. The extrinsic pathway is initiated when coagulation factor VII is exposed to and forms a complex with tissue factor (also known as tissue thromboplastin). The properties of factor VII as a zymogen are somewhat uncharacteristic since it can incorporate di-isopropylphosphofluoridate and exhibits up to 2% of the coagulant activity of factor VIIa. On binding of VIIa to the cell-bound tissue factor in the presence of calcium there is an estimated 10^7-fold enhancement of the activity of the enzyme compared to its free form[14]. Tissue factor is a glycoprotein with sequence similarity to the interferon receptor found in the membrane of certain cells. Its structure comprises a 200-residue extracellular domain, a 23-residue transmembrane domain and a 21-residue cytoplasmic tail. Normally absent from circulating blood cells and endothelium, tissue factor is selectively expressed in fibroblast-like cells surrounding the blood vessels. Factor VII will therefore only come into contact with tissue factor upon breakage of the wall of blood vessels, and such a pattern of distribution is thought to act as a haemostatic "envelope". Interestingly, atherosclerotic plaque contains many cells synthesizing tissue factor and rupture of plaque and exposure of tissue factor may lead to the thrombosis responsible for heart attack.

Factor VIIa is one of the vitamin K-dependent coagulation proteinases and it can activate factor X or factor IX, although it is now thought that the crucial formation of thrombin occurs via factor IX rather than directly via factor X. The reason for thinking this is that hereditary deficiency of factor IX and its cofactor, factor VIII, is associated with haemophilias B and A respectively. If haemostasis proceeded through the direct activation of factor X by the tissue factor–factor VIIa complex the involvement of factor IX would be unnecessary and there would be no resulting haemophilia. It is now known that activation of factor X by tissue factor–factor VIIa complex is inhibited by tissue factor pathway inhibitor (see below). Although direct activation of factor X may initiate coagulation, this reaction is quickly shut down and the clotting process is only maintained by activation of factor IX. Factor IXa and Xa exert their activities in intimate association with the non-enzymic cofactors factors Va and VIIIa respectively. Factors V and VIII are large glycoproteins with similarity to ceruloplasmin and are activated upon proteolytic cleavage of specific bonds by factor Xa or thrombin. The active cofactors are composed of subunits stabilized by divalent cations and they bind tightly to membranes containing negatively charged phospholipids, although a hydrophobic component is also considered to be important. Factors IXa and Xa also bind to membranes, primarily those containing negatively charged phospholipids, and, as mentioned above, the binding is dependent upon the presence of calcium ions (Figure 5). Assembly of the complex occurs by binding of cofactor and enzyme at different sites on the membrane surface (at rates approaching the theoretical, diffusion-limited rates for these reactions), followed by movement and rearrangements of the proteins on the surface until the complex is formed (again at rates comparable to those proposed for a two-dimensional diffusion process)[2].

Catalytically there are a number of changes conferred upon the reaction as it occurs in the complex. It has been shown that in the prothrombinase complex (factor Xa, factor Va, Ca^{2+}, cell membrane) which generates thrombin from prothrombin, a stoichiometric complex (1:1) of factor Xa and factor Va is formed, causing some alteration in the active site environment of factor Xa. Prothrombin also binds to factor Va and the cofactor therefore increases the probability of interaction between the enzyme and its substrate (an effect known as approximation). The overall effect is a calculated 300000-fold increase in the rate of the reactions. This increase is due to a decrease in the apparent K_m for prothrombin and an increase in the $k_{cat.}$ for the reaction. The apparent K_m is reduced from 100 μM to a value close to the concentration of prothrombin in blood (1 μM). Because of the influence of the concentration of the phospholipid upon K_m the suggestion has been made that the "K_m effect" is due to sequestering of prothrombin onto the membrane surface, thereby increasing the local concentration relative to the concentration of substrate in the bulk solution. The $k_{cat.}$ effect is ascribed to the influence of the cofactor on the activity of factor Xa, a phenomenon for which there is, as yet, no certain molecular explanation[2].

Localization: the formation of sites favouring coagulation on cell surfaces

As described above, an important feature of blood coagulation is the contribution of cell membranes, and there is some evidence to suggest that cell surfaces generate sites which favour coagulation (procoagulant sites). The activity of the vitamin K-dependent clotting factor is dependent on the negatively charged phospholipid phosphatidylserine which is assymetrically distributed in mammalian cell membranes, being preferentially localized in the inner leaflet. Certain conditions of cell stimulation may cause a disturbance in this distribution with increased exposure of phosphatidylserine in the outer leaflet of the cell membrane — this has been observed when platelets are treated with a mixture of thrombin and collagen[15]. Since blood will be exposed to collagen only when there is a break in a vessel, this is an appropriate combination of agonists to have such an effect on the platelet surface. Platelets may also undergo budding with release of microvesicles high in phosphatidylserine and rich also in factor Va. Other vessel cells may also be involved. Endothelial cells have been shown to have sites with high affinity for factor IXa which act to potentiate coagulation on the vessel wall. Increase in the procoagulant nature of endothelial cells has been shown to arise when the cells are damaged by physiological or extraneous insult (eg lack of oxygen or bacterial endotoxin). Recently, the generation of procoagulant sites has been shown to be intimately associated with integrin structures. The transition of the monocyte cell to macrophage occurs with the exposure of a number of integrin-like structures responsible for adherence of the cells to endothelial cells prior to diapedesis (migration into the extravascular milieu)[16]. One of these integrins, Mac-1, binds factor Xa and fibrinogen (fibrinogen is implicated in the cell–cell adhesion properties of the vascular cells) and this ligation would appear to be of importance in increasing the amount of tissue factor exposed by the cell.

SUMMARY OF THE ON-SWITCH

● Normal haemostasis depends on the rapid but localized generation of thrombin which clots soluble fibrinogen. The mechanisms and structures responsible for the formation of the enzyme are:

● the organization of highly specific proteinases which are converted to their active forms in a linked cascade by which a small initial stimulus is amplified;

● the domain structures of the coagulation proteinases and cofactors which help to localize and concentrate haemostatic factors at specific sites on the cell surface, enhancing each step in the coagulation cascade;

● the activation of essential cofactors by thrombin, causing a positive feedback mechanism favouring its own generation;

● the ability of vascular cells to generate specific sites on the membrane surface which accelerate the rate of each step of the coagulation cascade by up to 10^7-fold.

MECHANISMS RESPONSIBLE FOR THE LIMITATION OF BLOOD COAGULATION: THE MOLECULAR OFF-SWITCH

When the switch is thrown and clotting begins, what happens to turn off the formation of thrombin and fibrin? Three mechanisms are considered to limit and control blood coagulation.

Digestion of fibrin

The generation of fibrin monomer acts as a powerful catalyst of a proteolytic system (the fibrinolytic system) designed, seemingly, for the digestion of fibrin and removal of thrombi. The central enzyme of the fibrinolytic system is plasmin, a potent proteinase which normally exists in blood as a precursor, plasminogen[17]. Plasmin readily digests fibrinogen and fibrin and, in so doing, can dissolve whole blood clots (Figure 6a). Plasminogen is liable to activation by plasminogen activators of which there are three considered to be circulating in blood. The most important of these is tissue-type plasminogen activator, which is synthesized by and closely associated with the cells of the vessel wall (endothelial cells and possibly smooth muscle cells). On death the activator is released from the vessel wall and its presence was first observed as causing the liquefaction of the clotted blood in the veins of cadavers.

Activation of plasminogen by tissue type plasminogen activator is considerably enhanced in the presence of fibrin (10^2–10^3-fold) — plasminogen and tissue plasminogen activator have affinity for fibrin and together form a ternary complex — and significant amounts of plasmin are generated only in close association with a clot. Furthermore, plasmin is liable to inhibition by one of the serine proteinase inhibitors (serpins[18]) present in plasma, namely α_2-antiplasmin, a 67 kDa glycoprotein which is present in plasma at a concentration equivalent to about 60% of that of plasminogen. α_2-Antiplasmin combines with and inhibits plasmin at a rate approaching diffusional constraint (10^7 M^{-1} s^{-1}) through a mechanism involving the kringle domains of plasminogen. These kringle domains are also responsible for the affinity of plasmin for fibrin, and the inhibition of the enzyme α_2-antiplasmin is therefore impaired within the clot. As plasmin can act only within this environment, this ensures that only the clot is digested and not proteins in the circulating blood itself. The occurrence of free plasmin in blood would lead to bleeding by degrading fibrinogen and other clotting proteins. Because of its high potency in the presence of fibrin, tissue plasminogen activator has been developed for therapeutic removal of thrombi — thrombolytic therapy. As proteinases, the activators themselves are also liable to inhibition by other serpins, plasminogen activator inhibitors 1 and 2 (PAI-1 and PAI-2).

When plasmin is generated in the presence of fibrin it can directly inhibit fibrin polymerization by cleavage of a peptide from the N-terminus of the molecule. Loss of this peptide removes a polymerization zone and clotting is prevented. Plasmin can also dissolve well established crosslinked thrombi and therefore plays a part in wound healing.

Inhibition of clotting proteinases

The direct inhibition of the proteolytic activity of the clotting proteinases is another mechanism for the control of blood coagulation. Of the many proteinase inhibitors present in plasma the serpin antithrombin III is the most important, and heterozygous deficiency of the protein in individuals is associated with a life history of thrombotic disease. Since no individual with homozygous deficiency has yet been found, the absence of the inhibitor is probably incompatible with life. Antithrombin III is present

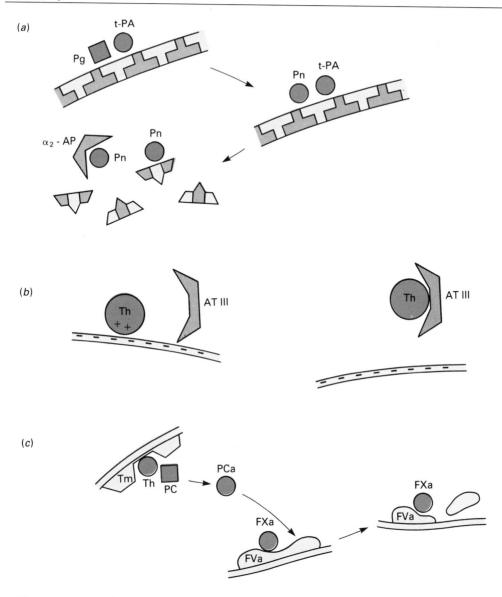

Figure 6. Mechanisms responsible for the limitation of blood clotting
(a) Plasminogen (Pg) is readily activated by tissue plasminogen activator (t-PA) to plasmin when it forms a ternary complex with the activator on fibrin strands within the clot. The plasmin (Pn) generated can digest fibrin polymer to soluble products but is rapidly inhibited by forming an inactive complex with the proteinase inhibitor α_2-antiplasmin (α_2-AP) as soon as it leaves the fibrin surface. These mechanisms localize the digestion to the clot. (b) In the presence of long heparan sulphate molecules present on the vessel wall, thrombin (Th) is rapidly inhibited by antithrombin III (ATIII) forming an irreversible complex. The heparan sulphate catalyses this reaction by binding the enzyme and the inhibitor, thereby increasing their affinity. (c) Protein C (PC) is activated by thrombin (Th) when it is in complex with a receptor, thrombomodulin (Tm), present on the surface of the cells of the blood vessel. The active enzyme protein Ca (PCa), in the presence of the cofactor, protein S, digests factor Va (FVa) or factor VIIIa thus destroying the cell-bound activation of prothrombin and factor X (FX) (see Figure 5).

in blood at a concentration of approximately 1.5 μM, equivalent to the concentration of prothrombin, and its target proteinases are considered to be thrombin (second order rate constant for association, $K_{ass.}$, 5.6×10^3 M^{-1} s^{-1}) and factor Xa ($K_{ass.}$ 4.3×10^2 M^{-1} s^{-1}). These association rates are fairly slow when compared to, for example, the inhibition of elastase by α_1-proteinase inhibitor (approaching 10^8 M^{-1} s^{-1}) and would be insufficient to control the activity of thrombin if more than about 0.1% of the blood prothrombin were activated. The association rates need to be increased for antithrombin III to act as an inhibitor of coagulation and this increase does occur in the presence of heparin[19]. Heparin, a highly sulphated glycosaminoglycan, has been used an an anti-coagulant for over 50 years and it causes increase in the association rate between anti-thrombin III and thrombin and factor Xa by 10^3–10^4 fold at a concentration of 10^{-9} M. The catalytic mechanism for this increase is due partly to conformational changes observed in the antithrombin III molecule upon binding of heparin and also by an approximating effect by which inhibitor and enzyme are bound to the same heparin molecule (Figure 6b). The tight binding of heparin to antithrombin III is highly specific and involves a unique pentasaccharide sequence. *In vivo*, heparin is located in mast cells but within the blood vessels it is considered that the antithrombotic activity of antithrombin III is in-creased by heparin-like heparan sulphate molecules present on the endothelium. Another serpin with properties very similar to antithrombin III is heparin cofactor II, which is only capable of inhibiting thrombin; this reaction is catalysed by heparin and dermatan sulphate.

As noted above, the tissue factor–factor VIIa complex is a very potent activator of coagulation. The activity of this complex can be inhibited by tissue factor pathway in-hibitor which is present on the endothelium and acts as a very powerful inhibitor of the first events of blood coagulation[20]. A glycoprotein of 38 kDa, it has three Kunitz-like domains linked in tandem. Kunitz domain 2 is a tight-binding inhibitor of factor Xa (K_d 10^{-9} M) and this initial complex binds to the tissue factor-factor VIIa complex.

Destruction of clotting cofactors and cell bound activity

A haemostatic factor for which a deficiency is associated with thrombotic disease is the vitamin K-dependent proteinase, protein C. Activated protein C (protein Ca) controls coagulation by destroying the activity of the cofactors factor Va and VIIIa. In this de-struction another vitamin K-dependent protein, protein S, acts as a cofactor (Figure 6c). Protein C is activated by a membrane bound Ca^{2+}-dependent complex composed of thrombin and thrombomodulin. Thrombomodulin functions as a cofactor for the reaction and is an integral membrane protein occurring on the endothelial cell surface[21]. The association of thrombin with thrombomodulin increases the rate of protein C activation 10^4-fold and also reduces the activity of thrombin towards fibrinogen and factor V. Thus as thrombin is generated it is inhibited directly and in so doing catalyses the anticoagulant events leading to the shutdown of its own generation.

SUMMARY OF THE OFF-SWITCH

● As thrombin is generated powerful anticoagulant mechanisms are immedi-ately initiated, limiting clot formation to the area of the wound. These mech-anisms are:

- the activation, by a fibrin-mediated mechanism, of plasminogen to plasmin, a proteinase which readily destroys the polymerization property of fibrin monomer and also can dissolve an established clot;

- the inhibition of thrombin and factor Xa by proteinase inhibitors, the rate of inhibition being accelerated by heparan sulphate present on the vascular wall;

- the inhibition of the factor VIIa–tissue factor complex by tissue factor pathway inhibitor;

- the thrombin-mediated formation of protein Ca, which destroys the essential cofactors factor Va and factor VIIIa.

MEDICAL ASPECTS OF BLOOD CLOTTING

Defects in the blood clotting system may be associated with excessive bleeding tendency (haemophilia) or tendency to thrombosis (thrombophilia). Haemophilia usually arises because of a lack of synthesis of one of the coagulation proteins, although it may also occur where there is synthesis of an abnormal, dysfunctional protein. The most widely known haemophilias are haemophilia A and B, arising from a lack of factor VIII and factor IX, respectively. A trivial cut in a haemophiliac may result in serious and prolonged bleeding often into the joints and deep tissues producing pain and growth abnormalities. The genes for these diseases are present on the X chromosome and for this reason the bleeding disorder occurs only in males who inherit the genetic abnormality from a female carrier in whom the disease is not seen. Haemophilia is treated by infusion at regular intervals of a factor IX or factor VIII concentrate prepared from blood donated for transfusion. It was necessary, until recently, to use concentrates since these proteins are present in such low concentration in plasma (especially factor VIII) that it was impossible to prepare pure material at sufficient yield to make it worthwhile. However, the use of concentrates has caused problems by inducing immunological inhibitors to the factor after prolonged therapy and more recently and tragically by infusing HIV virus with material prepared from the blood of AIDS victims, before the nature of the disease and its transmittance was fully appreciated. Today biotechnology has come to the rescue. Large-scale production of pure factor IX is now possible using monoclonal antibody columns to purify the protein. Recombinant factor VIII is also becoming available, an amazing achievement considering the large size of the molecule.

Thrombophilic patients (i.e. those with a life-long clinical history of thrombosis) are commoner than haemophilic patients but as yet it has only been possible to discover the abnormality responsible in 20% or so of cases. The commonest defects are deficiencies of protein C or antithrombin III and it is interesting to note that the disease will often be seen in heterozygotes in whom the level of the normal protein is 50%[22].

The description "thrombophilic" is not used for those individuals who may be destined for a thrombotic event, for example those with atherosclerotic disease (hardening of the arteries and high blood pressure) in which a fatal thrombus may develop in the blood vessels of the heart (heart attack) or brain (stroke). There is a very close relationship between levels of some of the blood clotting proteins and risk of death by the acute events of atherosclerotic disease[23]. In fact, epidemiological studies have demonstrated that high levels of fibrinogen and factor VIIa show a better degree of correlation with the risk of heart attack in middle-aged males than do cholesterol levels[24,25]. In the general population, without obvious atherosclerotic disease, the risk of thrombosis increases steadily as we get older but the event may only occur under conditions of stress or trauma. For example, 30% or so of patients over the age of 40 years undergoing major surgery will develop some degree of thrombosis in the deep veins of the leg. Such thrombosis can lead to sudden death if the thrombus increases in size and then dislodges blocking the vessels of the lungs (pulmonary embolism). Lung embolism is often responsible for the sudden death of surgical patients who are successfully operated and appear quite well but collapse when they get out of bed a few days later.

Thrombosis can be prevented in patients at high risk by the coumarin-type drugs (which alter the calcium-binding properties of the clotting factor) or by using heparin which activates the inhibitors of the clotting proteinases. Both these drugs, although powerful and effective, have their drawbacks and there is a continuing search for new drugs to supplement them. In the future it may be possible to use hirudin a powerful inhibitor of thrombin which is found as an anticoagulant in the saliva of leeches. Large amounts of hirudin can now be produced by recombinant technology. There is also increasing interest in the pharmaceutical industry in the design of small synthetic inhibitors of the coagulation proteases, especially thrombin.

When a thrombus does occur then how can it be removed rapidly and save lives? This is particularly of interest in the case of heart attack; intervention in the first hour or so after blockage occurs in the coronary arteries can perhaps prevent death and may also prevent further hypoxia-induced damage to the cardiac muscle, helping the long-term prognosis of the patient. For such treatment, what have come to be known as "clot busting drugs" have been introduced, in fact activators of plasminogen which induce digestion of the fibrin mesh of the thrombus[26]. One of the first biotechnological drugs to be introduced as a pharmaceutical was recombinant tissue plasminogen activator, advocated for single injection treatment of the heart attack victim in the ambulance. Other examples of such drugs are streptokinase, a plasminogen activator produced by Streptococcal bacteria, and a chemically inactivated complex of streptokinase and plasmin which is slowly activated upon infusion, a property which makes it more clot selective than streptokinase alone.

CONCLUSION

Blood coagulation is a defence mechanism to maintain the integrity of the blood circulation system. An insoluble fibrin plug is produced by a large pulse of thrombin generated specifically in the area of wound or injury. This pulse is initially generated from a primary stimulus which is amplified by a mechanism dependent on linked proteolytic reactions localized on sites present on the cell surface. These sites are created by the binding of activated cofactors formed by positive feedback loops. Concurrent to the generation of the coagulant activity, powerful anticoagulant mechanisms are initiated, ensuring that coagulation is limited to the wound area and shut down by these negative feedback loops.

REFERENCES

1. Jackson, C.M. & Nemerson, Y. (1980) Blood coagulation. *Annu. Rev. Biochem.* **49**, 765–811

2. Mann, K.G., Jenny, R.J. & Krishnaswamy, S. (1988) Cofactor proteins in the assembly and expression of blood clotting enzyme complexes. *Annu. Rev. Biochem.* **57**, 915–956

3. Furie, B. & Furie, B.C. (1988) The molecular basis of blood coagulation. *Cell* **53**, 508–518

4. Davie, E.W., Fujikawa, K. & Kisiel, W. (1991) The coagulation cascade: initiation, maintenance and regulation. *Biochemistry* **30**, 10363–10370

5. Coller, B.S. (1992) Platelets in cardiovascular thrombosis and thrombolysis, in *The Heart and Cardiovascular System* (Fozzard, H.A., Haber, E., Jennings, R.B., Katz, A.M. & Morgan, H.E., eds.), 2nd edn., pp. 219–273, Raven Press, New York

6. Scully, M.F. (1991) Plasminogen activator-dependent pericellular proteolysis: Annotation. *Br. J. Haematol.* **79**, 537–543

7. Patthy, L. (1985) Evolution of the proteases of blood coagulation and fibrinolysis by assembly from modules. *Cell* **41**, 657–663

8. Tulinsky, A. (1991). The structures of domains of blood proteins. *Thromb. Haemostasis* **66**, 16–31

9. Doolittle, R.F. (1984) Fibrinogen and fibrin. *Annu. Rev. Biochem.* **53**, 195–229

10. Fenton, J.W., II (1988) Thrombin bioregulatory functions. *Adv. Clin. Enzymol.* **6**, 186–193

11. Bode, W., Mayr, I., Baumann, U., Huber, R., Stone, S.R. & Hofsteenge, J. (1989) The refined 1.9 Å crystal structure of human α-thrombin: interaction with D-Phe-Pro-Arg chloromethylketone and significance of the Tyr-Pro-Pro-Trp insertion segment. *EMBO J.* **8**, 3467–3475

12. Vu, T.-K.H., Hung, D.T., Wheaton, V.I. & Coughlin, S.R. (1991) Molecular cloning of a functional thrombin receptor reveals a novel proteolytic mechanism of receptor activation. *Cell* **64**, 1057–1068

13. Griffin, J.H. & Bouma, B.N. (1987) The contact phase of coagulation, in *Haemostasis and Thrombosis* (Bloom, A.R. & Thomas, D.P., eds.), pp. 101–115, Churchill Livingstone, Edinburgh, London, Melbourne and New York

14. Edgington, T.S., Mackman, N., Brand, K. & Ruf, W. (1991) The structural biology of the expression and function of tissue factor. *Thromb. Haemostasis* **66**, 67–79

15. Bevers, E.M., Rosing, J. & Zwaal, R.F.A. (1985) Development of procoagulant binding sites on the platelet surface, in *Mechanisms of Stimulus Response Coupling in Platelets* (Westwick, J., Scully, M.F., McIntyre, D.E. & Kakkar, V.V., eds.), pp. 359–372, Plenum Press, New York and London

16. Arnaout, A.M.A. (1990) Structure and function of the leukocyte adhesion molecules: CD 11/CD 18. *Blood* **75**, 1037–1050

17. Bachmann, F. (1987) Fibrinolysis, in *Thrombosis and Haemostasis* (1987) (Verstraete, M., Vermylen, J., Lijnen, R. & Arnout, J., eds.), pp. 227–265, Leuven University Press

18. Carrell, R.W. & Boswell, R.B. (1986) Serpins: the superfamily of plasma serine proteinase inhibitors, in *Proteinase Inhibitors* (Barrett, A.J. & Salvesen, G., eds.), pp. 403–420, Elsevier, Amsterdam

19. Bjork, I., Olson, S.T. & Shore, J.D. (1989) Molecular mechanism of the accelerating effect of heparin on the reactions between antithrombin and clotting proteinases, in *Heparin* (Lane, D.A. & Lindahl, U., eds.), pp. 229–256, Edward Arnold, London, Melbourne and Auckland

20. Broze, G.J., Jr, Girard, T.J. & Novotny, W.F. (1991) Regulation of coagulation by a multivalent Kunitz-type inhibitor. *Biochemistry* **29**, 7539–7546

21. Esmon, C.T. (1989) The roles of protein C and thrombomodulin in the regulation of blood coagulation. *J. Biol. Chem.* **264**, 4743–4746

22. Cooper, D.N. (1991) The molecular genetics of familial venous thrombosis. *Blood* **5**, 55–70
23. Fuster, V., Badimon, L., Badimon, J.J. & Chesebro, J.H. (1992) Mechanisms of disease: the pathogenesis of coronary artery disease and the acute coronary syndromes. *New England J. Med.* **326**, 310–318
24. Meade, T.W. (1987) The epidemiology of haemostatic and other variables in coronary artery disease, in *Thrombosis and Haemostasis* (1987) (Verstraete, M., Vermylen, J., Lijnen, R. & Arnout, J., eds.), pp. 37–60, Leuven University Press
25. Cook, N.S. & Ubben, D. (1990) Fibrinogen as a major risk factor in cardiovascular disease. *Trends Pharm. Sci.* **11**, 444–451
26. Lijnen, H.R. & Collen, D. (1991) Towards the development of improved thrombolytic agents: Annotation. *Br. J. Haematol.* **77**, 261–266
27. Kyrle, P., Westwick, J., Scully, M.F., Lewis, G.P. & Kakkar, V.V. (1986) Changes in haemostatic variables at the bleeding site: influence of aspirin administration. *Thromb. Haemostasis* **57**, 62–66

3

Ubiquitin

R. John Mayer and Fergus J. Doherty

Department of Biochemistry, University of Nottingham Medical School,
Queen's Medical Centre, Nottingham NG7 2UH

INTRODUCTION

The small (76 residues) basic protein ubiquitin is as abundant as actin (in non-muscle cells) and is found in all eukaryotes so far examined, but is absent from prokaryotes. Ubiquitin appears to function through covalent attachment to other proteins, a novel post-translational modification that is turning out to have great biological importance. Around 50% or more of the ubiquitin in a cell is, in fact, coupled to other proteins. Goldstein first isolated and sequenced ubiquitin, which he identified as a lymphocyte differentiation promoting factor, in 1975. Ubiquitin was first found coupled to other proteins in 1977 when some nuclear proteins, histones H2A and H2B, were found to exist in "ubiquitinated" forms. An ATP-dependent Proteolysis Factor I isolated from extracts of rabbit reticulocytes by Hershko and co-workers, and which was covalently coupled to proteolytic substrates in the presence of ATP, was identified in 1980 by Haas and co-workers as being ubiquitin. Several roles for ubiquitin, therefore, have been determined so far, including the marking of intracellular proteins for degradation and modification of chromatin structure, but it now seems unclear if ubiquitin itself promotes lymphocyte differentiation.

STRUCTURE OF UBIQUITIN

Ubiquitin forms a compact globular structure with the C-terminus accessible and extending into the aqueous space (Figure 1). A short stretch of three and a half turns of α-helix together with a mixed β-sheet form a hydrophobic core. Nearly 87% of the polypeptide backbone is hydrogen-bonded, contributing to its remarkable stability to extremes of heat and pH and resistance to proteolysis. The accessible C-terminus is highly significant as it is the carboxyl group of the final glycine residue that forms

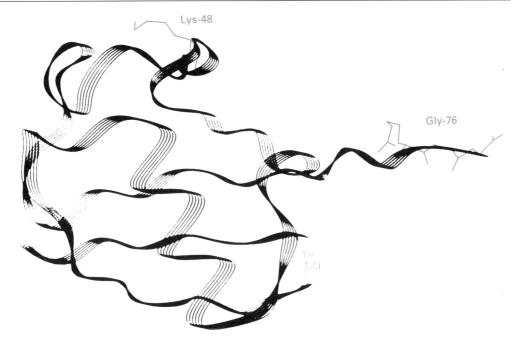

Figure 1. Ribbon representation of the three-dimensional structure of ubiquitin with wire frame representations of the acceptor lysine at position 48 and the C-terminal Arg-Gly-Gly superimposed

The C-terminal glycine (residue 76) is labelled. The ribbon represents the polypeptide backbone and clearly shows the region of α-helix running from bottom left to top right of the globular part of the molecule.

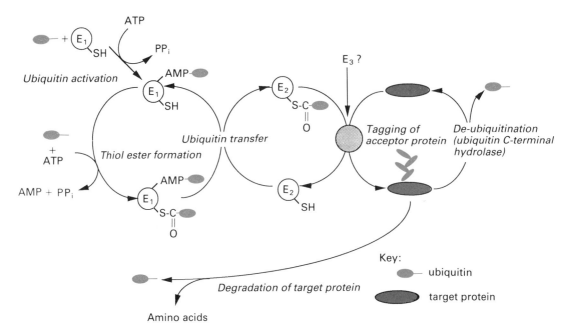

Figure 2. The ubiquitin cycle for protein degradation

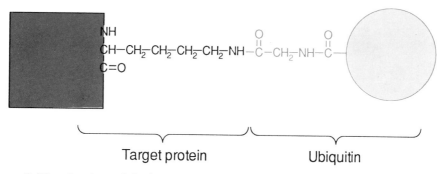

Figure 3. The structure of the isopeptide linkage between ubiquitin and a target protein

isopeptide bonds with amino groups of lysine residues in other proteins. This is the molecular event that underpins the function of ubiquitin.

ROLES OF UBIQUITIN

Ubiquitin and intracellular protein degradation

Protein degradation is important in regulating the intracellular concentration of proteins. Lysosomes are a major site for protein degradation; however, it was shown in reticulocytes (the circulating precursors of erythrocytes which lack lysosomes) that there is an ATP-dependent non-lysosomal proteolytic system involving ubiquitin. Ubiquitin is covalently attached to proteins in an ATP-dependent fashion, and these proteins are then rapidly degraded. Covalent attachment of ubiquitin therefore "tags" proteins in such a way that they are recognized and degraded by a non-lysosomal (cytosolic) protease. This mechanism is thought to be common to all eukaryotic cells.

The conjugation of ubiquitin to target proteins is a multi-step process (Figure 2). First, ubiquitin is activated to form ubiquitin-adenylate. The activated ubiquitin forms a complex with the activating enzyme, E_1 and then ubiquitin is transferred to another site on the E_1 enzyme to form a thiol ester between the terminal carboxyl group of ubiquitin and a sulphydryl (-SH), provided by a cysteine residue on the enzyme. Next the ubiquitin moiety is transferred to the thiol group of one of a family of E_2 or "ubiquitin-conjugating" (UBC) enzymes (Figure 2). Finally ubiquitin is ligated to the target protein by formation of a peptide bond between the terminal carboxyl group of ubiquitin (glycine) and an ε-amino group of a lysine residue on the target protein (Figure 3). Ligation of ubiquitin may require additional enzymes called E_3s which bind E_2–ubiquitin complexes together with a target protein, and catalyse the transfer of ubiquitin to the protein[1]. Processive addition of ubiquitin molecules, ligated to Lys-48 of the previous ubiquitin, can lead to a multi-ubiquitin chain of up to about 20 units (Figure 4) and, in addition, ubiquitin may be ligated to more than one acceptor lysine on a target protein. A mutant ubiquitin with a cysteine substituted for lysine at position 48 does not form multi-ubiquitin chains and does not support proteolysis; therefore, it is multi-ubiquitin chains that appear to target proteins for degradation[2].

Ubiquitinated proteins are degraded in the cytosol by a large molecular mass proteinase (>1000kDa) probably derived from the multicatalytic proteinase complex[3]. The multicatalytic proteinase complex is a multisubunit 19 S ring-shaped particle

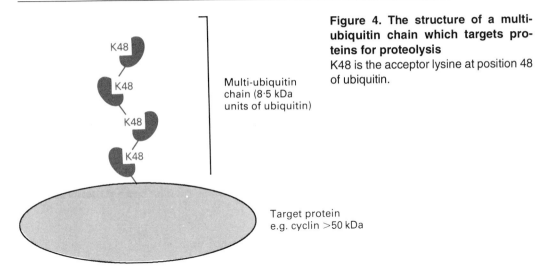

Figure 4. The structure of a multi-ubiquitin chain which targets proteins for proteolysis
K48 is the acceptor lysine at position 48 of ubiquitin.

Multi-ubiquitin chain (8·5 kDa units of ubiquitin)

Target protein e.g. cyclin >50 kDa

(prosome, or 19 S proteasome) which expresses multiple proteolytic activities and in the presence of ATP acquires additional subunits to form a larger structure (26 S proteasome) which selectively degrades proteins that are ubiquitinated. Proteolysis of ubiquitinated proteins by the 26 S proteasome is ATP-dependent and the target protein is probably degraded to small peptides or amino acids. The ubiquitin molecules are released intact from small peptides by isopeptidases known as ubiquitin C-terminal hydrolases. Other ubiquitin C-terminal hydrolase enzymes can "edit" multi-ubiquitinated proteins by removing ubiquitin units which may be necessary both for efficient degradation, by preventing the accumulation of massive conjugates, and as a "proof reading" mechanism to regenerate proteins ubiquitinated in error.

So what are the substrates for ubiquitin-mediated proteolysis? One group of proteins was indicated as proteolytic substrates by studies on a cell line (ts85) which harbours a temperature-sensitive mutation in the ubiquitin activating enzyme, E_1. At the non-permissive temperature short-lived intracellular proteins cannot be degraded in these cell lines, when the ubiquitin activating E_1 enzyme is inactive. The half-lives of intracellular proteins range from minutes to days and the proteins known to be short-lived are generally important regulatory enzymes. Some constitutively short-lived proteins, therefore, must express signals which target them for ubiquitination and rapid degradation. One such signal which has been identified is the nature of the amino acid found at the N-terminus of a protein[4] (Figure 5). Recognition of protein N-termini (the "N-end rule") is dependent on two E_3 enzymes. The physiological significance of the N-end rule is unclear, however, as cytosolic proteins do not express the destabilizing amino acids at the N-terminus. The identity of the residue expressed at the N-terminus of proteins is in fact determined by whether or not the initiator methionine is removed post-translationally by a methionine aminopeptidase. In fact, the methionine aminopeptidase does not remove the initiator methionine from eukaryotic cytosolic proteins to expose destabilizing amino acid residues at the N-terminus. In addition, most cytosolic proteins (approximately 80%) have N-termini blocked by N-acetylation and are not recognized by the N-end rule machinery. It

Figure 5. The N-end rule for protein degradation

(*a*) Methionine aminopeptidase cleaves off the initiator methionine to expose Xaa. If Xaa is a primary destabilizing residue (P: Arg, His, Lys, Phe, Leu, Trp, Tyr or Ile in yeast and reticulocytes, also Ala, Ser or Thr in reticulocytes) the protein is recognized by the N-end rule E_3, ubiquitinated and degraded (*b*). If Xaa is a secondary destabilizing residue (S: Glu or Asp, also Cys in reticulocytes) the protein is modified by the post-translational addition of an arginine (primary destabilizing) residue at the N-terminus (*c*) followed by binding to E_3, ubiquitination and degradation (*b*). If Xaa is a tertiary destabilizing residues (T: Gln or Asn) it is deaminated (*d*) to yield a secondary destabilizing residue (Glu or Asp) followed by addition of arginine at the N-terminus (*c*) then ubiquitinated and degraded (*a*).

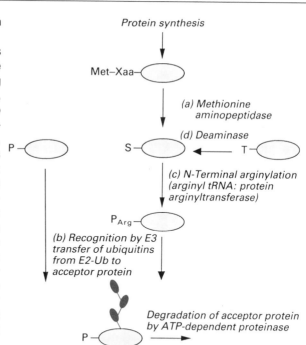

may be that the N-end rule permits the identification of mis-compartmentalized proteins, i.e. proteins destined for secretion or for intracellular compartments, that have inadvertently entered the cytosolic space. The search continues, therefore, for molecular signals downstream of the N-terminus that target constitutively short-lived intracellular proteins for ubiquitin-mediated degradation. One candidate signal may be an internal sequence such as the so-called "destruction box" of cyclin (see below). It is likely that more than one such internal signal exists, which in turn may imply the existence of more E_3 enzymes than are currently known.

It is important to remember that not all short-lived proteins are degraded by the ubiquitin-mediated pathway. However, individual proteins that have been identified as possible substrates of ubiquitin-mediated proteolysis are the far-red light absorbing form of phytochrome, the yeast α repressor, and cyclin (Table 1). The tumour repressor p53 is ubiquitinated and degraded by reticulocyte extract following association with a protein product (E6) of a human papillomavirus. In addition, oncogene products

Table 1. Natural substrates of ubiquitin-mediated proteolysis

Protein	Function	Reference
Cyclin	Cell cycle regulation	7
Phytochrome	Measurement of day length in plants	21
MAT α2 repressor	Transcriptional regulator in yeast	22
p53	Tumour repressor	23
myc, fos	Transcription factors	24

(*myc, fos*), which act as uncontrollable transcription factors, appear to be rapidly degraded by the ubiquitin system *in vitro*.

- Conjugation of ubiquitin to proteins is an ATP-dependent multi-step process.
- Ubiquitin is attached to target proteins by an isopeptide bond between the C-terminus of ubiquitin and the lysine ε-amino group on a target protein.
- Tagging of proteins with multi-ubiquitin chains targets them for degradation.
- In the presence of ATP the cytosolic 19 S proteasome (multicatalytic proteinase) assembles with additional subunits to form a 26 S proteasome that degrades ubiquitinated proteins in an ATP-dependent fashion.
- The cytosolic ubiquitin-mediated proteolytic system degrades short-lived proteins.

Ubiquitin and cell stress

Exposure of cells to elevated temperatures, e.g. 39 °C for mammalian cells, results in the "heat shock response". This involves the increased synthesis of a group of proteins designed to enable the cell to respond to the stress. One effect of heat stress is protein denaturation and some proteins are probably irreversibly damaged and must be removed. Ubiquitin conjugation appears to be involved in the removal of these damaged proteins[5]. Ubiquitin itself is a "heat shock protein" in that its expression is increased following exposure of cells to heat, presumably to enable rapid removal of damaged proteins. Ubiquitin-mediated proteolysis is also involved in the degradation of proteins made "abnormal" because they have incorporated amino acid analogues, an insult which invokes the synthesis of stress proteins. However, it is not known how the ubiquitin system recognizes damaged or abnormal proteins. Ubiquitination is somehow also involved in the repair of DNA (e.g. following ultraviolet-induced DNA damage) as one E_2 is the product of the yeast *rad6* gene, which is required for DNA repair[6]. Possibly ubiquitin-mediated proteolysis of nucleosome proteins allows access of the DNA repair enzymes to the damaged DNA. The *rad6* gene product is also specifically involved in the ubiquitination of proteins recognized by the N-end rule.

Ubiquitin and the cell cycle

Cyclins are proteins whose concentration in the cell oscillates with the cell cycle although their rate of synthesis is constant. The rapid degradation of cyclin which lowers its concentration and permits exit from mitosis (M to G_1) appears to be ubiquitin-mediated[7]. This is a good example of how protein degradation can determine the concentration of a regulatory protein. As indicated earlier the molecular signal within the cyclin molecule has been dubbed the "destruction box". In addition, a yeast E_2 enzyme is the product of the *cdc34* gene which controls cell cycle progression from G_1 to S. The substrate for this E_2 is unknown.

Programmed cell death and differentiation

Insect metamorphosis is a striking example of the importance of programmed cell death in development. Polyubiquitin gene expression increases markedly in the intersegmental muscles of the moth, *Manducca sexta*, following adult emergence from the pupa when these large muscles degenerate in response to a fall in ecdysteroids. Presumably, elevated ubiquitin-mediated proteolysis is part of the cell degeneration process[8]. Ubiquitin-mediated proteolysis is also involved in the loss of cell components as the reticulocyte matures into the erythrocyte and in plants a similar process of destruction of proteins and organelles occurs when vascular tissue is formed[9]. These three systems point to the importance of ubiquitin-mediated proteolysis in the cellular remodelling of differentiation, and in degeneration during programmed cell death.

- Cells respond to a variety of stresses by ubiquitination and degradation of damaged proteins.
- Ubiquitin-mediated proteolysis controls the concentration of cyclin which is involved in progression through the cell cycle.
- Loss of proteins and organelles during differentiation and programmed cell death involves ubiquitin-mediated proteolysis.

Ubiquitin and the lysosome

Ubiquitin–protein conjugates have been shown to be specifically enriched in the lysosomes of a variety of cell types relative to all other organelles[10,11]. Cells with E_1 inactivated at elevated temperatures (ts85 cells at 39 °C) cannot respond to stress by elevated degradation of proteins in their lysosome system[5]. These combined observations indicate that protein ubiquitination is somehow involved in lysosomal protein degradation. The mechanisms are not yet understood, but ubiquitination of proteins may be a signal for bulk sequestration of cytoplasm into a membrane-bordered autophagic vacuole which subsequently fuses with lysosomes (macroautophagy) or selective uptake of proteins from the cytoplasm into the lysosomal system (microautophagy). Another possibility is that ubiquitin is involved in lysosomal biogenesis.

Ubiquitin and histones

A ubiquitinated protein was first identified as a component of histone A24, a constituent of chromatin. A24 is in fact a ubiquitinated form of histone H2A (uH2A) and constitutes about 10% of this histone species while a smaller fraction (1%) of histone H2B is mono-ubiquitinated. uH2A is found as mono- and multi-ubiquitinated forms. The role of histone ubiquitination is somewhat controversial but it appears that ubiquitinated H2B is much more abundant in transcriptionally active chromatin. Ubiquitination of H2B, which probably occurs when the nucleosome is unfolded during transcription, may impede refolding and permit repeated rounds of transcription[12]. The role of H2A ubiquitination is less clear. H2B ubiquitination apparently does not target histones for degradation as this requires the presence of multi-ubiquitin chains.

Ubiquitin and the cell surface

A number of cell-surface proteins are known to exist in ubiquitinated forms. These are the lymphocyte homing receptor, platelet derived growth factor receptor, growth hormone receptor and the choline uptake transporters of neurones. In the case of the lymphocyte homing receptor ubiquitin is thought to be found on the outer, extra-cellular face of the protein. It is not known if modification by ubiquitination alters the turnover of these proteins but the presence of ubiquitin at the cell surface poses the question: how does it get there? The recent identification of a gene encoding a yeast membrane-bound E_2 enzyme could mean that proteins may be ubiquitinated as they pass into, or through, a membrane.

Ubiquitin and viruses

Proteins sharing sequence homology with ubiquitin are coded in several viral genomes e.g. bovine viral diarrhoea viruses and baculovirus. It is well known that in the course of viral infection gene fragments can be integrated into viral genomes, e.g. retroviral incorporation of genes for oncoproteins. Fragments of genes for ubi-quitin-like proteins may have been incorporated into viral nucleic acids for several reasons. A ubiquitin-like protein which was sufficiently like ubiquitin to be activated and conjugated but not act as a degradation signal could act as a poison of the ubiquitin system, preventing the system from eliminating viral proteins from the cell. Some viral particles also contain ubiquitin conjugated to viral proteins, e.g. tobacco mosaic virus[13]. The purpose of this ubiquitination is unknown, but ubiquitin may have a biosynthetic chaperoning role as in ribosome biogenesis (see below). Alterna-tively, ubiquitination of proteins in viral particles may be evidence of an attempt by the cell to eliminate viral particles by protein degradation. Recently, it has been dem-onstrated that an active E_2 which is similar to a yeast E_2 (RAD6) is coded in the genome of a DNA virus (African swine fever virus). The role of this enzyme in viral infection and replication is unknown.

Ubiquitin as a "chaperone"

Some ubiquitin genes direct the synthesis of ubiquitin fusion proteins with the C-terminus of ubiquitin followed directly by the N-terminus of another protein. The C-terminal extensions of these ubiquitin fusion proteins are ribosomal protein sub-units[14]. The ubiquitin moiety of the fusion protein is cleaved off before the ribosomal protein subunit is incorporated into the ribosome. The purpose of this unusual gene structure is unknown but the ubiquitin moiety may in some way aid in the trans-location and incorporation of the protein subunit into the ribosome. In this way ubiquitin may serve as a "molecular chaperone", as, indeed, do other heat shock proteins.

Ubiquitin cross-reactive protein

The connection between ubiquitin-mediated metabolism and cell defence (for which we use the term cytoprotection) is emphasized by the finding of a ubiquitin cross-reactive protein (UCRP) whose concentration increases in cells treated with inter-feron[15]. Interferon is produced by virally infected cells and secreted to activate surrounding cells to defend themselves from subsequent viral infection. The UCRP is highly homologous, but not identical, to a dimer of ubiquitin and occurs in cell extracts as UCRP–protein conjugates as well as free UCRP. The function of UCRP conjugation is currently unknown but recent evidence shows that it is slowly secreted from lymphocytes and monocytes. UCRP may be a novel cytokine.

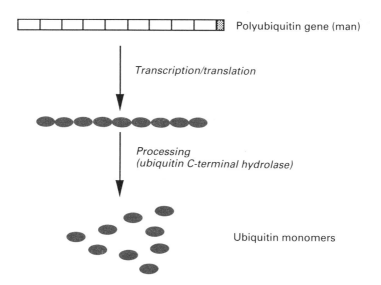

Figure 6. Expression of polyubiquitin genes

Ubiquitin and disease

Ubiquitin–protein conjugates are found in intracellular filamentous inclusion bodies in degenerating neurones. These accumulations of insoluble proteins include the neurofibrillary tangles of Alzheimer's disease, Lewy bodies in Parkinson's disease and diffuse Lewy body disease, and inclusion bodies in motor neurone disease[16]. Cytoskeletal proteins are found in a number of these inclusions. Filamentous inclusions with ubiquitin–protein conjugates are not only found in disorders of the nervous system but occur in other diseased tissues, e.g. in hepatocytes in alcoholic liver disease (Mallory bodies). There is, therefore, a family of ubiquitin-filament disorders. The reason for the formation of the inclusions containing the ubiquitinated proteins is not fully understood but it must in some way represent an attempt to remove or isolate abnormal proteins. Interestingly Epstein–Barr virus-transformed lymphoblastoid cells contain a "cocoon" which consists of lysosomes containing a viral membrane protein, ubiquitinated proteins and the heat shock protein 70, surrounded by intermediate filaments[17]. This may be a similar process to that found in the diseases to isolate "foreign" or abnormal proteins.

- Protein modification by ubiquitination is not solely for targeting for degradation.
- Ubiquitinated forms of nuclear histones may be involved in regulating transcription.
- Ubiquitinated proteins can be found at the cell surface.
- Some viruses have genes for ubiquitin-like proteins and ubiquitin conjugating enzymes.
- Some important human diseases are characterized by the presence of intracellular insoluble protein deposits which contain ubiquitinated proteins.

GENES OF THE UBIQUITIN SYSTEM

Ubiquitin has the most conserved amino acid sequence known, varying in only three amino acids from yeast to man. This may result from the multifunctional nature of ubiquitin with the molecule recognized by a number of different systems. Ubiquitin may have evolved originally as a modifier of histones and then been recruited for other purposes. Ubiquitin is coded as a set of fusion proteins by a family of genes. The genes are of two types, the polyubiquitin genes, and other genes which code for fusion proteins consisting of ubiquitin with a C-terminal extension corresponding to ribosomal protein subunits.

There are several genes in each eukaryotic organism which code for polyubiquitins. In yeast there is one polyubiquitin gene coding for five ubiquitin molecules[18], in man there are two polyubiquitin genes coding for polyproteins containing nine and three ubiquitin molecules[19]. In some lower organisms, for example parasites, polyubiquitin genes coding for up to 40 ubiquitin repeats have been demonstrated. The translated polyubiquitins are cut by ubiquitin C-terminal hydrolases to release the ubiquitin molecules which can then be used in ubiquitin-mediated metabolic processes (Figure 6). The polyubiquitin genes have heat-shock promoters and are therefore switched on in cells subjected to some form of stress (e.g. in heat stressed cells such as mammalian cells at temperatures of 39–43 °C) or injury. The increased levels of ubiquitin, rapidly produced by expression of genes encoding multiple copies of ubiquitin, must help cells cope with the stress (see above).

In yeast there are three genes which code for ubiquitin fusion proteins while in man there are probably two of these genes (Figure 7). The genes have promoters resembling those found for ribosomal protein genes. The reason for the encoding of one or two subunits by ubiquitin-fusion genes may be that the ubiquitin moiety acts as a chaperone (see above). Ubiquitin-ribosomal protein genes may also serve in some way to couple protein synthesis and protein degradation. Finally it has been suggested that the ubiquitin encoding region may stabilize the mRNA.

Already nine genes encoding ubiquitin conjugating enzymes (E$_2$s) have been identified in yeast. The number of ubiquitin carrier enzymes may reach the number of

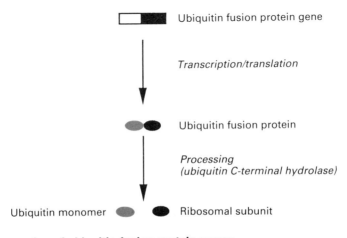

Figure 7. Expression of ubiquitin fusion protein genes

protein kinases! Some of these gene products have been demonstrated to be essential for ubiquitin-dependent proteolysis, others may be involved in other forms of ubiquitination, e.g. of histones. Deletion of all three yeast E_2 genes involved in ubiquitin-mediated proteolysis is lethal to the cell[20].

In summary, ubiquitin plays a vital role in a number of key biological processes, including the degradation of intracellular proteins and the modification of chromatin, by covalently tagging proteins. Possibly other roles for ubiquitin remain to be discovered. Key areas of current research include identification of the enzymes of the ubiquitin pathway and the substrates *in vivo* of ubiquitin-mediated proteolysis. In addition, the molecular signals in proteins which determine whether or not they are ubiquitinated remain to be discovered.

- Ubiquitin is coded for by multiple genes.

- Polyubiquitin genes code for head-to-tail repeats of ubiquitin, the resulting polyprotein being processed to ubiquitin monomers by a specific protease.

- Ubiquitin fusion genes code for fusion proteins of ubiquitin with a C-terminal ribosomal protein extension which are post-translationally cleaved.

- Polyubiquitin genes have a heat shock promoter.

- Ubiquitin conjugating enzymes are coded for by multiple genes.

REFERENCES

1. Reiss, Y. & Hershko, A. (1990) Affinity purification of ubiquitin-protein ligase on immobilized protein substrates – evidence for the existence of separate NH_2-terminal binding sites on a single enzyme. *J. Biol. Chem.* **265**, 3685–3690

2. Chau, V., *et al.* (1989) A multiubiquitin chain is confined to specific lysine in a targeted short-lived protein. *Science* **243**, 1576–1583

3. Driscoll, J. & Goldberg, A.L. (1990) The proteasome (multicatalytic protease) is a component of the 1500-kDa proteolytic complex which degrades ubiquitin-conjugated proteins. *J. Biol. Chem.* **265**, 4789–4792

4. Bachmair, A., Finley, D. & Varshavsky, A. (1986) *In vivo* half-life of a protein is a function of its amino-terminal residue. *Science* **234**, 179–186

5. Gropper, R., *et al.* (1991) The ubiquitin-activating enzyme, E_1, is required for stress-induced lysosomal degradation of cellular proteins. *J. Biol. Chem.* **266**, 3602–3610

6. Jentsch, S., McGrath, J. & Varshavsky, A. (1987) The yeast DNA repair gene RAD6 encodes a ubiquitin-conjugating enzyme. *Nature (London)* **329**, 131–134

7. Glotzer, M., Murray, A.W. & Kirschner, M.W. (1991) Cyclin is degraded by the ubiquitin pathway. *Nature (London)* **349**, 132–138

8. Schwartz, L.M., Myer, A., Kosz, L., Engelstein, M. & Maier, C. (1990) Activation of polyubiquitin gene expression during developmentally programmed cell death. *Neuron* **5**, 411–419

9. Bachmair, A., Becker, F., Masterson, R.V. & Schell, J. (1990) Perturbation of the ubiquitin system causes leaf curling, vascular tissue alterations and necrotic lesions in a higher plant. *EMBO J.* **9**, 4543–4549

10. Doherty, F.J., *et al.* (1989) Ubiquitin-protein conjugates accumulate in the lysosomal system of fibroblasts treated with cysteine proteinase inhibitors. *Biochem. J.* **263**, 47–55

11. Laszlo, L., Doherty, F.J., Osborn, N.U. & Mayer, R.J. (1990) Ubiquitinated protein conjugates are specifically enriched in the lysosomal system of fibroblasts. *FEBS Lett.* **261**, 365–368

12. Davie, J.R. & Murphy, L.C. (1990) Level of ubiquitinated histone H_2B in chromatin is coupled to ongoing transcription. *Biochemistry* **29**, 4752–4757
13. Dunigan, D.D., Dietzgen, R.G., Schoelz, J.E. & Zaitlin, M. (1988) Tobacco mosaic virus particles contain ubiquitinated coat protein subunits. *Virology* **165**, 310–312
14. Redman, K.L. & Rechsteiner, M. (1988) Extended reading frame of a ubiquitin gene encodes a stable, conserved, basic protein. *J. Biol. Chem.* **263**, 4926–4931
15. Haas, A.L., Ahrens, P., Bright, P.M. & Ankel, H. (1987) Interferon induces a 15-kilodalton protein exhibiting marked homology to ubiquitin. *J. Biol. Chem.* **262**, 11315–11323
16. Lowe, J., *et al.* (1988) Ubiquitin is a common factor in intermediate filament inclusion bodies of diverse type in man, including those of Parkinson's disease, Pick's disease, and Alzheimer's disease, as well as Rosenthal fibres in cerebellar astrocytomas, cytoplasmic bodies in muscle, and Mallory bodies in alcoholic liver disease. *J. Pathol.* **155**, 9–15
17. Laszlo, L., *et al.* (1991) The latent membrane protein-1 in Epstein–Barr virus-transformed lymphoblastoid cells is found with ubiquitin–protein conjugates and heat-shock protein 70 in lysosomes oriented around the microtubule organising centre. *J. Pathol.* **164**, 203–214
18. Ozkaynak, E., Finley, D. & Varshavsky, A. (1984) The yeast ubiquitin gene: head-to-tail repeats encoding a polyubiquitin precursor protein. *Nature (London)* **312**, 663–666
19. Wiborg, O., *et al.* (1985) The human ubiquitin multigene family: some genes contain multiple directly repeated ubiquitin coding sequences. *EMBO J.* **4**, 755–759
20. Seufert, W. & Jentsch, S. (1990) Ubiquitin-conjugating enzymes Ubc4 and Ubc5 mediate selective degradation of short-lived and abnormal proteins. *EMBO J.* **9**, 543–550
21. Shanklin, J., Jabben, M. & Vierstra, R.D. (1989) Partial purification and peptide mapping of ubiquitin phytochrome conjugates from oat. *Biochemistry* **28**, 6028-6034
22. Hochstrasser, M. & Varshavsky, A. (1990) *In vivo* degradation of a transcriptional regulator – the yeast alpha-2 repressor. *Cell* **61**, 697–708
23. Scheffner, M., Werness, B.A., Hulbregtse, J.M., Levine, A.J. & Howley, P.M. (1990) The E6 oncoprotein encoded by human papillomavirus types 16 and 18 promotes the degradation of p53. *Cell* **63**, 1129–1136
24. Ciechanover, A., *et al.* (1991) Degradation of nuclear oncoproteins by the ubiquitin system *in vitro. Proc. Natl. Acad. Sci. U.S.A.* **88**, 139–143

FURTHER READING

Rechsteiner, M.C. (1988) *Ubiquitin* 1–346, Plenum Press, New York

Ciechanover, A. & Schwartz, A.L. (1989) How are substrates recognized by the ubiquitin-mediated proteolytic system? *Trends Biochem. Sci.* **14**, 483–488

Jentsch, S., Seufert, W., Sommer, T. & Reins, H.A. (1990) Ubiquitin-conjugating enzymes – novel regulators of eukaryotic cells. *Trends Biochem. Sci.* **15**, 195–198

Mayer, R.J., Arnold, J., Laszlo, L. Landon, M. & Lowe, J. (1991) Ubiquitin in health and disease. *Biochim. Biophys. Acta* **1089**, 141–157

Hershko, A. (1991) The ubiquitin pathway for protein degradation. *Trends Biochem. Sci.* **16**, 265–268

Jentsch, S., Seufert, W. & Hauser, H.P. (1991) Genetic analysis of the ubiquitin system. *Biochim. Biophys. Acta* **1089**, 127–139

4

The collagen superfamily – diverse structures and assemblies

David J.S. Hulmes

Department of Biochemistry, University of Edinburgh, Hugh Robson Building, George Square, Edinburgh EH8 9XD, U.K.

INTRODUCTION

Collagens, like most structural proteins, display an enormous capacity for diversity. Since the discovery of collagen polymorphism less than 25 years ago, the number of recognized vertebrate collagen types has grown to its present total of 14. Additionally, there are proteins with collagen-like motifs that are not called collagens. The remarkable feature of this collagen superfamily is its varied repertoire of molecular structures and patterns of supramolecular assembly. Here we explore the mechanisms that give rise to this diversity, how assembly is controlled and how the collagens are designed to fulfil a wide range of biological functions.

What is a collagen? The usual definition is a protein with three polypeptide chains where each chain contains at least one stretch of the repeating amino acid sequence (Gly-Xaa-Yaa)$_n$ and Xaa and Yaa can be any amino acid (often proline and hydroxyproline, respectively). The three chains, which may or may not be identical, combine together via their (Gly-Xaa-Yaa)$_n$ sequences to form a collagen triple helix[1]. The triple-helical regions are identified experimentally by their susceptibility to bacterial collagenase and resistance to proteinases such as trypsin or pepsin, or by their characteristic X-ray diffraction pattern. Though all collagens necessarily contain at least one triple-helical region, not all proteins with (Gly-Xaa-Yaa)$_n$ sequences are called collagens. These "non-collagen-collagens" are known by the properties of their non-triple-helical regions. But the distinction between collagens and non-collagens is becoming increasingly blurred as sequence similarities are revealed and as new functions are discovered for the non-triple-helical regions. It is therefore more appropriate to consider the collagen superfamily (Figures 1 and 2), which includes all proteins with a collagen triple helix.

Vertebrate collagens can be classified (Table 1) on the basis of their size and self-assembly[1,2]. The classical collagen fibril, observed by electron microscopy as a roughly cylindrical structure (diameter 20–500 nm) with a characteristic 64–67 nm (D) repeating banding pattern, is the form of assembly favoured by collagens I, II, III, V and XI; these are the fibrillar (D-staggered) collagens. The non-fibrillar collagens (IV, VII, VIII, IX, X, XII, XIII, XIV) are more difficult to classify, but distinct kinds of assembly occur and natural subdivisions are emerging with the burgeoning of new sequence data.

Table 1. Vertebrate collagens

Type	α chains	Most common molecular form	Tissue distribution
I	$\alpha1(I)$, $\alpha2(I)$	$[\alpha1(I)]_2\alpha2(I)$	Most connective tissues, e.g. bone, tendon, skin, lung, cornea, sclera, vascular system
II	$\alpha1(II)$	$[\alpha1(II)]_3$	Cartilage, vitreous humour, embryonic cornea
III	$\alpha1(III)$	$[\alpha1(III)]_3$	Extensible connective tissues, e.g. skin, lung, vascular system
IV	$\alpha1(IV)$, $\alpha2(IV)$, $\alpha3(IV)$, $\alpha4(IV)$, $\alpha5(IV)$	$[\alpha1(IV)]_2\alpha2(IV)$	Basement membranes
V	$\alpha1(V)$, $\alpha2(V)$, $\alpha3(V)$	$[\alpha1(V)]_2\alpha2(V)$	Tissues containing collagen I, quantitatively minor component
VI	$\alpha1(VI)$, $\alpha2(VI)$, $\alpha3(VI)$	$\alpha1(VI)\alpha2(VI)\alpha3(VI)$	Most connective tissues, including cartilage
VII	$\alpha1(VII)$	$[\alpha1(VII)]_3$	Basement-membrane-associated anchoring fibrils
VIII	$\alpha1(VIII)$, $\alpha2(VIII)$	$[\alpha1(VIII)]_2\alpha2(VIII)$?	Product of endothelial and various tumour cell lines
IX	$\alpha1(IX)$, $\alpha2(IX)$, $\alpha3(IX)$	$\alpha1(IX)\alpha2(IX)\alpha3(IX)$	Tissues containing collagen II, quantitatively minor component
X	$\alpha1(X)$	$[\alpha1(X)]_3$	Hypertrophic zone of cartilage
XI	$\alpha1(XI)$, $\alpha2(XI)$, $\alpha3(XI)$*	$\alpha1(XI)\alpha2(XI)\alpha3(XI)$	Tissues containing collagen II, quantitatively minor component
XII	$\alpha1(XII)$	$[\alpha1(XII)]_3$	Tissues containing collagen I, quantitatively minor component
XIII	$\alpha1(XIII)$	$[\alpha1(XIII)]_3$?	Quantiaively minor collagen, found e.g. in skin, intestine
XIV	$\alpha1(XIV)$	$[\alpha1(XIV)]_3$?	Tissues containing collagen I, quantitatively minor component

*Closely related to $\alpha1(II)$

Type VI collagen forms fibrils, but not the D-staggered variety, and so is in a category of its own.

COLLAGEN NOMENCLATURE

Collagen (or procollagen) molecules are composed of three α (or pro-α) chains. Different chains within a single molecule are distinguished by arabic numerals, while different collagen types are distinguished by roman numerals in parentheses. For example, collagen I has the chain composition $[\alpha1(I)]_2\alpha2(I)$. pC-collagen (see below) is procollagen with the N-propeptides removed; similarly, pN-collagen is procollagen with the C-propeptides removed.

FIBRILLAR (D-STAGGERED) COLLAGENS (I, II, III, V, XI)

Molecular assembly

All the fibrillar (D-staggered) collagens are initially synthesised in precursor form, as procollagens (Figures 1 and 2). The production of a procollagen molecule (consisting of three pro-α chains) is a complex biosynthetic process, where diversity in the protein product can result from a large number of post-transcriptional and post-translational events. Most information has come from studies on type I procollagen biosynthesis[3,4], which probably serves as a paradigm for the rest of the collagen family.

Fibrillar collagen genes[5,6] consist mostly of non-coding regions. For example, the gene for the human pro-α1(I) chain is 18 kbp long, while the polypeptide chain contains 1423 amino acids, so only about 24% of the gene codes for protein. Large introns separate a total of 51 exons, the positions and sizes of which are remarkably conserved in the genes for procollagens I, II and III. Within the region of the gene that codes for the $(Gly-Xaa-Yaa)_n$ sequences in the mature collagen molecule, all the exons sizes are 54 bp, or a related multiple of 9 bp, which suggests that fibrillar collagens have evolved from an ancestral 54 bp unit. In contrast, the exons that code for the N- and C-terminal propeptide domains are relatively large and unrelated to 54 bp. Most N-propeptide domains comprise a cysteine-rich globular region, a short $(Gly-Xaa-Yaa)_n$ segment and a connecting sequence (Figure 1). The cysteine-rich region is absent from the pro-α2(I) chain, and in pro-α1(II) the exon that codes for this region can be spliced in or out to produce tissue-specific alternatively spliced variants[7]. Further examples of alternative splicing have been found in some of the non-fibrillar collagens (see below).

Procollagens are subject to a large number of post-translational modifications[3,4]. Following synthesis of prepro-α chains, cleavage of the signal peptides and insertion into the cisternae of the rough endoplasmic reticulum, pro-α chains are subject to inter-chain disulphide cross-linking, prolyl 4-hydroxylation (and, to a lesser extent, prolyl 3-hydroxylation), lysyl hydroxylation, hydroxylysine O-linked glycosylation in the triple-helical region, and N-linked glycosylation in the C-propeptide region. Concomitant with these enzymic events are chain registration (the association of pro-α chains in the correct stoichiometry) and triple-helix assembly. Disulphide bonding accompanies chain registration and precedes triple helix formation. Remarkably, the β subunit of the tetrameric enzyme prolyl 4-hydroxylase is identical to protein di-

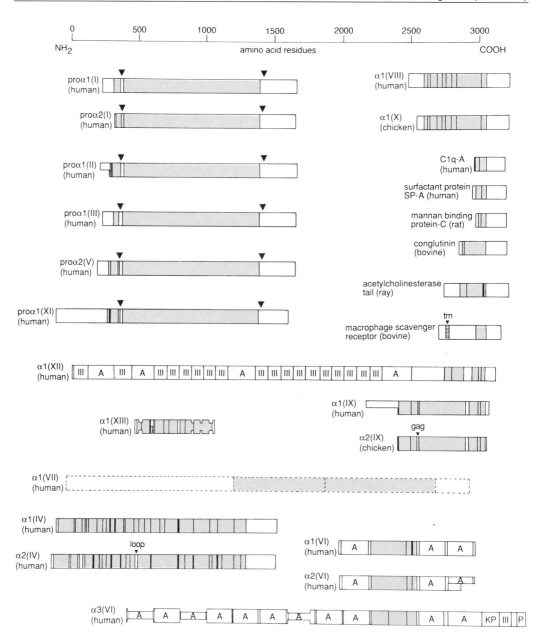

Figure 1. Amino acid sequences

Representative collagen α (or pro-α) chains, and non-collagen "collagen" chains, are shown diagrammatically with (Gly-Xaa-Yaa)$_n$ sequences in red. N-Termini are to the left, and short interruptions in the (Gly-Xaa-Yaa)$_n$ sequence are indicated by vertical black lines. Larger non-triple-helical regions are shown as white boxes which are subdivided, where appropriate, into discrete structural domains (see key). Alternative splicing can lead to the absence of certain sequences (indicated by boxes of half height) or the expression of alternative sequences (half-height boxes at the same axial position). Propeptide cleavage sites are indicated by large arrows (in the case of procollagens V and XI these are putative cleavage sites deduced from sulphide isomerase, the enzyme that catalyses disulphide exchange in the molecular assembly of collagens and other secretory proteins[8]. Procollagen molecular assembly begins by association of C-propeptide domains, then the triple helix folds in a zipper-like manner from the C- to N-terminus, at a rate determined by *cis/trans* isomerization of proline- (or hydroxyproline-) containing peptide bonds[9]. The prolyl hydroxylases, lysyl hydroxylase and the O-linked glycosyltransferases are active only on a non-triple-helical substrate. Because of this, any delay in triple-helix formation can lead to over-modified (i.e. over-hydroxylated and over-glycosylated) pro-α chains, as observed in a number of heritable disorders of collagen[3,4]. Mutant procollagen molecules are usually unstable at physiological temperatures so, because secretion requires a triple-helical conformation, they are degraded intracellularly. Following transport through the Golgi system, procollagen molecules are packaged in secretory vesicles where, unlike the D-staggered assembly that occurs in collagen fibrils, procollagen molecules are aligned in register (Figure 2). The form of procollagen aggregation is strongly dependent on environmental factors since, in contrast to the non-staggered assembly that occurs in secretory vesicles, purified procollagen I at high concentrations assembles in D-staggered array[10].

Fibril assembly

Newly synthesized procollagen molecules are secreted into the extracellular matrix, where they encounter a number of processing enzymes and undergo fibril formation[4] and cross-linking[11]. The enzymes are procollagen N-proteinases (removal of N-propeptides), procollagen C-proteinase (removal of C-propeptides) and lysyl oxidase (initiation of cross-linking). All the procollagen proteinases isolated so far are neutral, calcium-dependent metalloproteinases[12]. There are different N-proteinases for procollagens I and III, which allows for independent control of processing and assembly in tissues containing both collagen types (e.g. lung, skin, vascular system). Type I procollagen N-proteinase is a high-molecular-mass complex (500 kDa). Procollagen C-proteinase is active in monomeric (90 kDa) form, although activity is increased several-fold by a C-proteinase enhancer protein[13]. Lysyl oxidase[14] initiates cross-linking in collagens (and elastin) by oxidative deamination of certain lysine and hydroxylysine residues located in the short N- and C-terminal non-triple-helical regions (called telopeptides) that remain after removal of the procollagen propeptides. The resulting aldehyde derivatives react spontaneously with lysines or hydroxylysines in the triple-helical regions on adjacent collagen molecules to form Schiff base cross-links. With time, and depending on whether it is lysine or hydroxylysine that is initially oxidized, bifunctional cross-links undergo further intra- and inter-molecular reactions to form a variety of mature, trifunctional cross-links. Cross-link diversity accounts

sequence comparisons). The diagrams are based on complete cDNA or protein sequences, except for collagen VII (approximate dimensions represented by dotted outline), for which only partial sequence information is available. All chains are drawn to the same scale. Original sources are either cited in the text or can be found in the cited reviews. Key: tm, transmembrane domain; gag, glycosaminoglycan attachment site; loop, internal disulphide-bonded loop; III, fibronectin type III domain; A, von Willebrand factor A domain; KP, salivary protein-like domain; P, aprotinin-like domain.

for some of the major differences between skeletal and non-skeletal connective tissues[11]. The significance of hydroxylysyl glycosylation is not clear, though it may be important in fibril formation[15].

It is relatively recently that collagen fibril formation has been studied *in vitro* by enzymic processing of procollagen precursors[4,16]. Before these studies, the reconstitution of fibrils following neutralization and warming of acidic collagen solutions[17] had been used as a model for fibril assembly *in vivo*. Collagen fibrils reconstituted *in vitro* are similar to naturally occuring fibrils. Their D-periodic banding patterns are identical, though it is usually more difficult to reproduce fibril diameters. When fibril formation is studied *in vitro* by enzymic removal of C-propeptides from pC-collagen, the kinetics of assembly are similar, although the rate of enzyme processing provides an additional tier of control, and fibril diameters are relatively large. Much attention has been given to the problem of fibril diameter regulation, and a variety of possible mechanisms have been proposed (see below).

A useful way to think of assembly is in terms of nucleation and growth, where the rate-limiting step is the formation of a critical nucleus from a small number of molecules. Once the nucleus is formed, growth proceeds rapidly by accretion of monomers. Such a model is widely used in studies of other entropy-driven self-assembly systems[18] (e.g. actin, tubulin, flagellin), but these are in some ways simpler than the collagen system as the assembled polymers are usually of a well-defined structure and growth only occurs in length. In such systems, large numbers of nuclei are associated with small average lengths of polymer, for a fixed total concentration of protein. With collagen fibril assembly, there are possible changes in both length and diameter, and the two are not necessarily related. Changes in the number of nuclei may reduce, increase or have no effect on fibril diameter, depending on changes in fibril length. In most studies of fibril growth, either *in vitro* or *in vivo*, only diameters have been determined. Recent three-dimensional reconstructions from serial sections of embryonic chick tendon[19] have identified individual fibrils (diameter about 40 nm) of length about 10 μm, and the fibrils have pointed and blunt ends. Similar observations have been made *in vitro*, where the pointed end corresponds to the N-terminus[16]. The way in which molecules add to growing collagen fibrils is not well understood. It has been suggested, mainly on the basis of fibril reconstitution studies *in vitro*, that fibrils assemble from microfibrillar substructures, although the evidence for microfibrils in intact fibrils is inconclusive[20,21]. Figure 2 summarizes much of the available data on the molecular packing in collagen fibrils.

It is now well documented that most collagen fibrils are composed of two or more different collagen types[1]. For example, the fibrils in adult cornea, whose small diameter (20 nm) and regular organization are essential for corneal transparency, consist of collagens I and V. In cartilage, the fibrils contain collagens II, IX and XI, while in skin fibrils collagens I and III are present. Heterotypic interactions between different collagen types have been proposed as a possible mechanism for fibril diameter limitation, and there is evidence to support this from reconstitution studies *in vitro*[22]. Collagens also interact with other components of the extracellular matrix and, amongst these, proteoglycans are likely to play a major role in fibril assembly. A large number of proteoglycans have now been characterized and some of the small proteoglycans (e.g. decorin) bind to collagen and may regulate fibril diameter[23,24]. Collagens V and

XI are quantitatively minor collagens, associated with collagens I and II respectively, that appear to be located in the core of heterotypic collagen fibrils[1] and retain their N-propeptide extensions (Figure 1). The fibril-forming collagens differ in their relative rates of N- and C-terminal procollagen processing, and limited persistence of pN-collagen is associated with small diameter fibrils[25]. Persistence of large amounts of pN-collagen is the basis of a heritable disorder of connective tissue called dermato-sparaxis (fragile skin), originally seen in cattle and sheep (but also in humans; P.H. Byers, personal communication), where deficient N-propeptide processing leads to highly abnormal collagen fibrils (Figure 3). Deficient processing can be due to a defective enzyme (or ancilliary protein), as in dermatosparaxis, or to a mutation in the procollagen substrate, as in Ehlers–Danlos Syndrome Type VII, a related human heritable disorder in which exon skipping leads to loss of the N-proteinase cleavage site[4]. Recent experiments have shown that the abnormal fibrils can be reproduced *in vitro* by controlling the extent of N-terminal procollagen processing (Figure 3; see also[25]).

Effects of mutations

The phenotypic consequences of mutations in collagen genes are now quite well understood[3,4]. Osteogenesis imperfecta (brittle bone disease) is a group of heritable disorders (with phenotypes from mild to lethal) that affect about 1 in 10000 individuals, with clinical features such as short stature, bone deformity, hearing loss, and blue sclerae. Over 80 osteogenesis imperfecta mutations have been characterized to date in the genes for procollagen I. Single base changes in the codons for glycine account for most of the mutations, which reflects the strict stereochemical requirement for glycine at every third position in the collagen triple helix[1]. Substitutions of glycine for cysteine are readily detected at the protein level, since cysteines are normally absent from collagen I and disulphide bonding of cysteines in two adjacent mutant chains shows up clearly by electrophoresis. Glycine substitutions for most other single-base-change-related amino acids have also been found. As well as point mutations in the coding regions, single base changes in the intervening, non-coding sequences can affect splicing and lead to exon skipping, and deletions and insertions have also been characterized[3,4].

There is a correlation between the position of the mutation in the collagen I molecule and the severity of the osteogenesis imperfecta phenotype[3]. Usually mutations near the C-terminus are lethal, while the phenotype is milder as the N-terminus is approached. The reason for this is clear from studies on the assembly of the procollagen molecule, where triple-helix formation proceeds in a zipper-like manner from the C- to N-terminus[9]. Any mutation that interferes with folding can potentially destabilize the triple helix, and regions N-terminal to the mutation are likely to spend longer in a non-triple-helical conformation and therefore undergo more extensive post-translational modification (as the hydroxylases and O-linked glycosyltransferases are active only on non-triple-helical chains). The extent of such "overmodification" is therefore inversely correlated with distance of the mutation from the C-terminus. The severity of the phenotype also depends on the nature of the amino acid that is substituted for glycine[3,4]. Most mutations lower the melting temperature by just a few degrees, but this is enough to unfold the triple helix at physiological temperatures and therefore

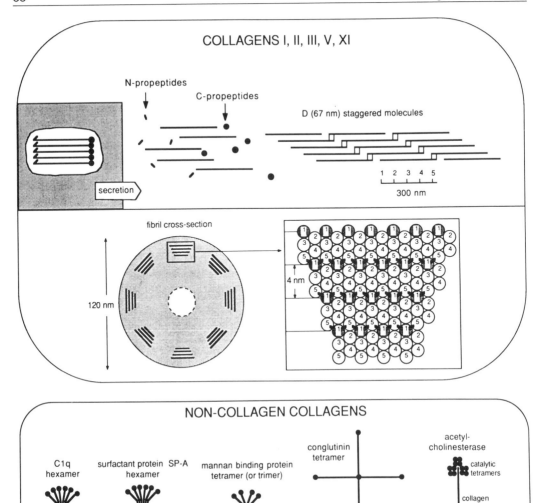

Figure 2 (above and right). Molecules and assemblies
Diagrams of individual molecules (shown with N-termini to the left) are based on electron microscopic images of rotary shadowed preparations. Lines represent triple-helical regions, and non-triple-helical regions are shown as filled circles (or tridentate structures). Note the use of different scales. Collagens I, II, III, V and XI: Procollagen secretion, processing, D-staggered collagen assembly and cross-links (short vertical lines) are indicated. The rates and extents of processing differ between different collagen types. The fibril in cross-section shows crystalline domains, as observed[21] in fibrils of collagen I. Within a crystalline domain, a likely molecular packing arrangement is shown in the enlarged view, where each circle represents a molecule in cross-section, and the numbers refer to relative axial locations (see scale in top part of figure). Cross-links between adjacent molecules are shown by short thick lines. There is evidence for a concentric, helical[45] organization around the fibril core (lighter shading), which in the case of heterotypic fibrils may consist of collagen V or XI. Crystalline molecular packing has only been observed in the outer regions of large diameter collagen I fibrils, in general the lateral packing (i.e. perpendicular to the fibril axis) is likely to be much less regular. Collagen IV: The lower part of the figure (note different scale) is from[27], with permission. Collagen VI: Beaded filaments can

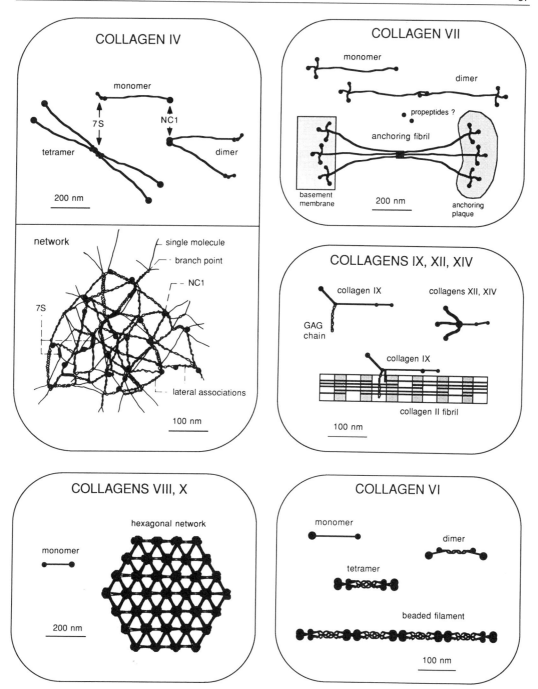

associate side-by-side to form 110 nm periodic fibrils (see Figure 3). Collagen VII: Propeptide processing appears to occur during maturation of dimers into anchoring fibrils. Collagens VIII, X: Nodes are likely to be connected by bundles of molecules, though the number of molecules in each bundle is unknown. Collagens IX, XII, XIV: The location of a cross-link between collagens II and IX is shown by a short vertical line. Non-collagen "collagens": Note the relatively small scale.

prevent secretion. Unfolded molecules accumulate intracellularly and are slowly degraded, so less extracellular matrix is produced. The effect of mutations is compounded by the phenomenon of "procollagen suicide", where, because of the multisubunit nature of the procollagen molecule, the number of mutant chains has a disproportionate effect on the number of mutant molecules. When mutant procollagen molecules are secreted, the potential for disruption is even greater, since the presence of a relatively small proportion of mutant molecules can markedly affect fibril assembly. This effect was recently seen in a human heterozygote with a Gly → Cys mutation in the α1(I) chain, where the introduction of cysteine leads to kinked molecules that copolymerize with normal molecules to form highly branched fibrils[26].

NON-FIBRILLAR COLLAGENS

While fibrillar collagens make up a closely related group within the collagen superfamily, there is a wide diversity in the structure and assembly of the non-fibrillar collagens, and in their genes[5,6]. It is convenient to think of the following three subgroups: basement membrane collagens, short-chain collagens and fibril-associated collagens.

Basement membrane collagens (IV, VII)

The term "basement membrane" is something of a misnomer[27]. Basement membranes (see Figure 3) are thin (30–40 nm), rather amorphous structures that are found in the extracellular matrix adjacent to plasma membranes and which line epithelial and endothelial cell layers, at the interface with mesenchyme, or surround particular cell types (e.g. muscle, fat, nerve). Their functions are to provide support for cell layers, to serve as molecular sieves (e.g. in the glomerulus), to act as selective barriers to the passage of inflammatory and tumour cells and to provide substrates for cell adhesion, growth and differentiation. The principal macromolecular components of basement membranes are collagens IV (there are multiple forms), laminins, nidogen (entactin) and heparan sulphate proteoglycans. Collagen VII occurs in close proximity to basement membranes and so is also included in this category, though as yet no sequence similarity to collagen IV has been detected[28].

The structure of the collagen IV molecule differs is several ways from the fibrillar (D-staggered) collagens (Figures 1 and 2). First, it is more appropriate to compare collagen IV with the procollagen precursors of the fibrillar collagens, as collagen IV does not undergo proteolytic processing. The central triple helical region of collagen IV is about 25% longer than the fibrillar collagens, and unlike the fibrillar collagens, it is interrupted at several positions by short non-triple helical sequences (Figure 1). In human collagen IV, for example, there are over 20 interruptions, most of which coincide in both the α1(IV) and α2(IV) chains. The positions of the interruptions are conserved in several species, and they correspond to sites of increased molecular flexibility, as seen in the electron microscope after rotary shadowing[27]. In the N-terminal region of the molecule (called 7 S, from its sedimentation constant), there is a further triple-helical region that is separated from the main triple helix by a kink. At the position of the kink, the α2(IV) chain is one residue shorter than α1(IV). A further difference from the fibrillar collagens is the presence in collagen IV of N-linked glycosylation in the triple-helical regions (both 7 S and the main helix) and the much greater extent of lysyl hydroxylation and hydroxylysine glycosylation. At the C-terminus of the collagen IV molecule there is a large globular domain (called NC1, for Non-Collagenous) which shows little sequence homology with the C-propeptide of the fibrillar collagens and consists of two homologous internal domains.

The assembly of collagen IV involves intermolecular interactions that are mediated by the 7 S and NC1 domains (Figure 2). The NC1 domains on two adjacent molecules become covalently linked, by disulphide exchange, to form a dimer. The 7 S domains also associate, in interactions that are stabilized by both disulphide and lysine-derived cross-links, to form a tetramer in which two pairs of anti-parallel molecules overlap by about 30 nm. When both 7 S and NC1 interactions are present, collagen IV can form an open network structure. An additional level of assembly involves lateral associations of the main triple-helical regions in interactions that are also mediated by NC1. The details of these lateral associations are poorly understood, but, from high resolution studies of basement membranes following removal of non-collagenous components, there is evidence for helical structures that appear to result from molecules being entwined together[27]. These lateral interactions must be important in functions such as molecular sieving, but their assembly poses the topological problem of how molecules can become entwined whilst being attached to other molecules at both ends. One end must be free for molecules to wrap round each other, and both 7 S-linked tetramers or NC1-linked dimers are intermediates in assembly[29]. Many of the features of the organization of collagen IV can be reproduced *in vitro*, in thermally activated reconstitution experiments similar to those used in studies on the fibrillar collagens, and detailed interactions with the other basement membrane components, laminin, nidogen and heparan sulphate proteoglycans, have been identified[27].

Collagen VII is found close to the basement membrane zone beneath stratified squamous epithelia (e.g. at the dermal–epidermal junction in skin) where it occurs in the form of anchoring fibrils, symmetric structures (about 750 nm long) that link the basement membrane to anchoring plaques in the underlying extracellular matrix (Figures 2 and 3). Collagen VII has the largest triple-helical region of the vertebrate collagens (Figure 1), and it has a unique form of self-assembly. The molecule consists of a triple-helical region about 420 nm long, with a small globular region at one end (probably the C-terminus[28]), which appears to be proteolytically removed during intermolecular assembly, and a large (50 nm) tridentate structure at the other end (Figure 2). As with collagen IV, the triple-helical region of collagen VII contains several non-helical interruptions. Molecules associate to form anti-parallel dimers, stabilized by disulphide bonds and with a small overlap at the small globular end, and then dimers associate in register to form the anchoring fibril bundle (Figure 2). The function of collagen VII in strengthening the dermal–epidermal junction is apparent in the recessive form of an heritable blistering disorder, dystrophic epidermolysis bullosa, where extremely mild mechanical stress leads to separation of the epithelial basement membrane from the underlying matrix and hence severe blister formation. In this disorder, histological and biochemical studies show that both anchoring fibrils and collagen VII are absent[1].

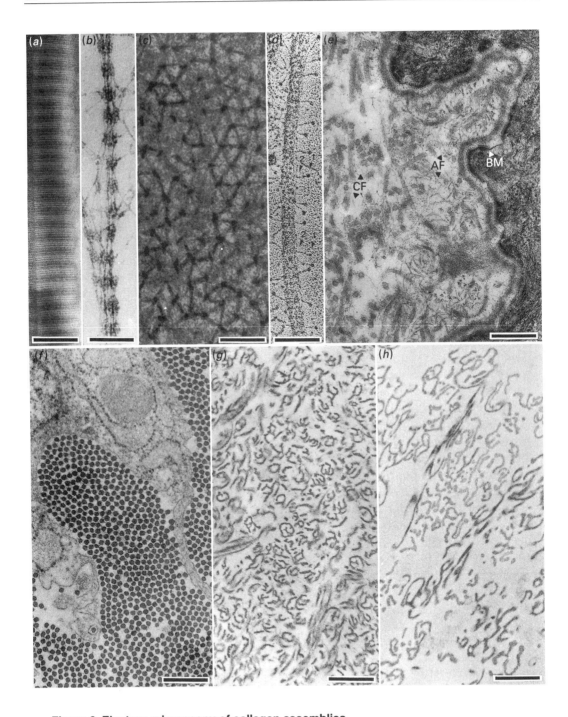

Figure 3. Electron microscopy of collagen assemblies

(a) Collagen I fibrils in mouse tail tendon. (b) Collagen VI fibrils in chick embryo tendon cell cultures, from[46] with permission. (c) Collagen VIII in Descemet's membrane of bovine cornea, courtesy of Dr R. Bruns. (d) Collagen IX molecules on the surface of a collagen II fibril, from[47] with permission. The N-terminal globular (1) and triple-helical (2) domains of collagen IX are

Short chain collagens (VIII, X)

The short chain collagens (types VIII and X) are related in structure and assembly, though their distributions and biological functions are distinct.

Collagen VIII was originally found in cultures of vascular and corneal endothelial cells[2], but it is also a product of both normal and malignant cells of non-endothelial origin. The molecule (Figures 1 and 2) consists of a triple helical region of length about 135 nm (less than half the length of a fibrillar collagen molecule) with a globular region at the C-terminus and a smaller globular region at the N-terminus. Both $\alpha1$(VIII) and $\alpha2$(VIII) chains have been characterized[1], and their genes are unusual as the entire triple-helical and C-terminal globular regions are coded by a single large exon. As with all non-fibrillar collagens, the triple helices of collagen VIII are interrupted by imperfections in the $(Gly-Xaa-Yaa)_n$ repeat and the interruptions are in the same positions in the $\alpha1$(VIII) and $\alpha2$(VIII) chains. The assembled form of collagen VIII has been observed in a specialized basement membrane of the corneal endothelium called Descemet's membrane (Figure 3). Here the macromolecular organization is in the form of an open hexagonal network where the separation between nodes is somewhat less than the length of a collagen VIII molecule (Figure 2). The function of such a highly organized form of basement membrane is not understood but is may be important in withstanding intraocular pressure. There is no proteolytic processing during the assembly of collagen VIII, but some forms of collagen VIII can undergo rapid intermolecular (acid-labile) cross-linking[2].

Collagen X is the most specialized of the collagens, and is the product of hypertrophic chondrocytes in the deep-calcifying zone of cartilage. All long bones grow in length by the process of endochondral ossification[30], in which chondrocytes in the cartilaginous growth plate become hypertrophic prior to calcification of cartilage and its eventual replacement by bone. Chondrocytes from the hypertrophic zone can be grown in culture, where they secrete large amounts of collagen X. This short-chain collagen is strongly similar to collagen VIII, especially in the triple-helical and C-terminal globular regions (Figure 1). Both collagen types have the same appearance in the electron microscope, the positions of the non-triple helical interuptions are conserved, and the resemblance extends to the gene level[1]. The assembled form of collagen X *in vivo* is in pericellular filamentous networks and also there is some colocalization with collagen II fibrils. Assembly *in vitro* has recently been studied by turbidity and rotary shadowing, where hexagonal networks remarkably similar to Descemet's membrane were observed. Possible functions of collagen X include forming a scaffolding during replacement of cartilage by bone or, by analogy with collagen VIII, in guidance of endothelial cells during angiogenesis[30].

indicated. (*e*) Dermal–epidermal junction of human skin showing basement membrane (BM), anchoring fibrils (AF) and collagen fibrils (CF), courtesy of Ms J. Spencer. (*f*) Collagen I fibrils in cross-section, from chick embryo tendon, courtesy of Mrs C. Cummings. Note fibril bundles in close association with the cell surface and in membraneous invaginations (arrowed). (*g*) Skin from child with dermatosparaxis, showing abnormal collagen fibrils in cross-section, courtesy of Drs K.A. Holbrook and L.T. Smith. (*h*) Abnormal collagen I fibrils in cross-section, produced *in vitro* by limited N-terminal procollagen processing, from[25] with permission. Scale bars: (*a*), (*b*), (*c*) 200 nm; (*d*) 100 nm; (*e*), (*f*), (*g*), (*h*) 400 nm.

Fibril-associated (FACIT) collagens (IX, XII, XIV)

FACIT collagens[31] (Fibril-**A**ssociated **C**ollagens with **I**nterrupted **T**riple helices) do not form fibrils on their own, but attach specifically to the surface of pre-existing (fibrillar collagen) fibrils. The original FACIT collagen was type IX, a quantitatively minor collagen that colocalizes with collagen II.

The collagen IX molecule is relatively short, about 200 nm in length, and consists of three triple-helical domains (with some single residue imperfections) interspersed between four non-collagenous (NC) domains (Figure 1). The three chains are distinct, they are connected by disulphide bonds and there is a large globular region (NC4) at the N-terminus of cartilage $\alpha1(I)$. The most unusual feature of collagen IX is that it often occurs as a proteoglycan[32], where a single dermatan sulphate glycosamino-glycan chain is covalently attached to the NC3 domain of the $\alpha2(IX)$ chain. The attachment site contains a five-residue loop that gives rise to the observed kink in the molecule at this position. Collagen IX molecules decorate the surface of collagen II fibrils (Figure 3), and the isolation of pyridinoline and other lysine-derived cross-links between the α chains of collagens II and IX shows the interaction to be highly specific (Figure 2; see also[1]). The globular NC4 domain of $\alpha1(IX)$ has an estimated pI of about 10, so this domain is likely to bind to the acidic glycosaminoglycan chains of the cartilage proteoglycan matrix, while the glycosaminoglycan chain of collagen IX may stabilize its interaction with collagen II. Collagen IX is also expressed, although transiently, in developing chick cornea, but the molecule lacks a globular NC4 domain[31]. This is not due to proteolytic processing, but to the use of a second transcription start site that introduces an alternative first exon otherwise found in an intron. The corneal transcript is thus shorter than the cartilage form, but is otherwise identical apart from a short sequence coded by the alternative first exon. The functional significance of the alternative forms of collagen IX is not understood. A further source of diversity is the length of the covalently attached glycosaminoglycan chain, which in the collagen IX of vitreous humour is about ten times longer than the cartilage form[31].

Collagen XII is a quantitatively minor collagen found in dense, collagen I-containing connective tissues such as tendons and ligaments[33]. The molecule consists of three $\alpha1(XII)$ chains, where each chain has two triple-helical domains (Figure 1), one of which is strongly homologous to collagen IX, and there is also a collagen IX NC4-like globular domain. These similarities suggest that collagen XII may also be a fibril-associated collagen, though specific fibril associations have not yet been observed. But thereafter the analogy with collagen IX breaks down, as collagen XII turns out to be the largest of vertebrate collagens known to date. The bulk of the sequence consists of several globular domains and the collagenous domains make up less than 8% of the entire molecule, so it is debatable whether this protein should really be called a collagen at all! The globular domains are of two types: there are 18 fibronectin type III domains and four von Willebrand factor type A domains. By rotary shadowing, the non-triple-helical regions of collagen XII appear as a large tridentate structure that is attached, via the collagen IX-like NC4 domain, to the small collagenous region (Figure 2).

Collagen XIV, also found in skin and tendon, is the latest of the FACIT collagens to be recognized. It is most like collagen XII in the triple-helical and nearby non-collagenous domains[34], though any further homology with collagen XII must await further sequencing.

Collagen XIII

Collagen XIII has so far only been characterized at the cDNA and genomic levels[35]. It is not homologous to any other collagen. The remarkable feature of collagen XIII is its complex pattern of alternative splicing (Figure 1). Recent analysis of the surprisingly large (at least 140 kbp) gene has shown that seven exons are involved in the production of at least 11 alternative transcripts. Alternative splicing has also been seen in collagens II, VI and IX, but so far collagen XIII is the only one to involve the triple-helical regions. The function of collagen XIII is unknown: *in situ* hybridization has shown mRNA expression in skin and intestinal mucosa, but information at the protein level is eagerly awaited.

COLLAGEN VI

Collagen VI is a relatively ubiquitous fibrillar collagen[1], but the fibrils are unlike those of the classical D-staggered collagens. Instead, the fibrils have a 110 nm periodicity and consist of filamentous microfibrils, with regularly spaced pairs of "beads on a string" (Figure 3).

The three chains of collagen VI are mostly non-collagenous (Figure 1). Each has an interrupted triple-helical region about one-third the length of a fibrillar (D-staggered) collagen, and this is flanked by relatively large N- and C-terminal globular domains. The non-collagenous domains make up about 65% of the $\alpha1(VI)$ and $\alpha2(VI)$ chains, while in $\alpha3(VI)$ they account for about 90% of the mass. Collagen VI is extensively glycosylated, with hydroxylysine O-linked carbohydrate and N-linked oligosaccharides in the triple-helical region.The non-collagenous domains are made up mostly of repeating von Willebrand factor A modules, a widely occuring protein motif[36] (also found in collagen XII) which in collagen VI undergoes alternative splicing[37,38]. The A domain is generally involved in heterophilic interactions involving cell surface receptors, serum proteins and matrix proteins, and in particular binding to collagen I. A further feature of collagen VI is the large number (11) of Arg-Gly-Asp (RGD) sequences throughout the triple helix. The RGD sequence is required for binding of several extracellular proteins (e.g. fibronectin, collagens, fibrinogen) to cell-surface integrin receptors[39]. Most collagen α chains contain only one or two such sequences per α chain, so the presence of such a large number of RGD sequences in collagen VI, along with the collagen I binding A domains, suggests a possible bridging role between cells and the extracellular matrix.

The assembly of collagen VI has been studied in cell culture and in tissue extracts[1]. Molecules first form anti-parallel dimers, where the molecules are overlapped for most of their length, and then dimers associate side-by-side to form tetramers (Figure 2). Tetramers are stabilized by an extensive intermolecular disulphide bond network that appears to form intracellularly (there are no lysine-derived cross-links in collagen VI), and microfibril assembly occurs extracellularly, though in close association with the cell surface, by end-to-end linkage of tetramers. Fibril formation by association of microfibrils has not been studied in detail, neither has it yet been possible to reproduce the assembly process *in vitro* from purified collagen VI molecules.

NON-COLLAGEN "COLLAGENS"

As the list of vertebrate collagens continues to grow, the non-collagenous domains have become increasingly significant, though their functions are poorly understood. In contrast, the non-collagen "collagens" (complement component C1q[40], mammalian lectins[41], acetylcholinesterase tail subunit[42], macrophage scavenger receptor[43]) are identified by the ligand-binding activities of their non-collagenous domains (Figures 1 and 2).

The C1q molecule has six identical subunits, where each subunit consists of a three-chain collagen-like structure with a large C-terminal globular domain. Subunits associate with their collagenous domains in register, and a kink in the triple helix [caused by interruptions in the (Gly-Xaa-Yaa)$_n$ repeat] allows the globular regions to spread out in a "flower bouquet" structure. The globular domain binds to immunoglobulin Fc (which triggers C1 activation) but it is also strongly homologous to the C-terminal non-collagenous domains of the short-chain collagens (VIII and X). The conserved residues may define a framework for chain alignment prior to triple-helix formation[40].

Several mammalian lectins contain collagenous domains[41]. The pulmonary surfactant protein SP-A, for example, has a C1q-like flower bouquet structure, but the C-terminal domain is unrelated to C1q and has a sequence that conforms to the Ca^{2+}-dependent (C-type) lectin domain. Assembly of the disulphide-bonded SP-A hexamer is a cell-mediated event, and like other collagens the C-terminal region is essential for chain alignment and initiation of helix formation[44]. Other collagen-like proteins with C-type lectin domains include mannan-binding protein, which is similar to C1q and SP-A but appears mainly as trimeric or tetrameric "bouquets", and conglutinin, a spoke-like tetramer of subunits with relatively long collagen-like regions[41].

The asymmetric form of acetylcholinesterase consists of up to three catalytic tetramers attached to a collagen-like tail subunit[42], which serves to anchor the enzyme at the neuromuscular junction by interactions with basement membrane proteoglycans. Finally, a continuous collagen triple-helix and α-helical coiled coil link the ligand-binding domain to the membrane-spanning domain of the macrophage scavenger receptor[43].

CONCLUSIONS

The collagen superfamily displays an impressive repertoire of structures and assemblies. Fibrillar (D-staggered) collagens, synthesized as procollagens, have continuous triple helices for specific D-staggered assembly. Complex interactions occur between different collagen types and with other extracellular matrix components. Interrupted triple helices are features of the non-fibrillar collagens, and the interruptions are important for increased molecular flexibility or formation of rigid kinks. The large terminal non-triple-helical regions can be crucial for both molecular and supramolecular assembly. In some collagens, the non-triple-helical regions dominate triple-helical regions, but the functions of these large globular domains are poorly understood. Similarities between some collagens and non-collagen "collagens" may point to underlying functional similarities. Collagen diversity is further increased by alternative splicing, and tissue-specific expression of splicing variants points to subtle functional differences. There is presently much structural information on the collagen superfamily, although our understanding of function remains sketchy. It is likely that current advances in molecular biology will rapidly begin to close this gap in the very near future.

I am grateful to numerous colleagues, but in particular to Romaine Bruns, John Chapman, Jerome Gross, Andrew Miller and Darwin Prockop. The author is a Wellcome Trust Research Leave Fellow.

REFERENCES

In general, original papers are only cited when they are not cited in the reviews included here.

1. van der Rest, M. & Garrone, R. (1991) Collagen family of proteins. *FASEB J.* **5**, 2814–2823
2. Mayne, R. & Burgeson, R.E. (eds.) (1987) *Structure and Function of Collagen Types*, pp. 1–317, Academic Press, Orlando
3. Byers, P.H. (1990) Brittle bones – fragile molecules: disorders of collagen gene structure and expression. *Trends Genet.* **6**, 293–300
4. Kuivaniemi, H., Tromp, G. & Prockop, D.J. (1991) Mutations in collagen genes: causes of rare and some common diseases in humans. *FASEB J.* **5**, 2052–2060
5. Sandell, L.J. & Boyd, C.D. (eds.) (1990) *Extracellular Matrix Genes*, pp. 1–270, Academic Press, Orlando
6. Vuorio, E. & de Crombugghe, B. (1990) The family of collagen genes. *Annu. Rev. Biochem.* **59**, 837–872
7. Sandell, L.J., Morris, N., Robbins, J.R. & Goldring, M. (1991) Alternatively spliced type II procollagen messenger RNAs define distinctive sub-populations of cells during vertebrate development – differential expression of the amino propeptide. *J. Cell Biol.* **114**, 1307–1319
8. Freedman, R.B., Bulleid, N.J., Hawkins, H.C. & Paver, J.L. (1989) Role of protein disulphide isomerase in the expression of native proteins. *Biochem. Soc. Symp.* **55**, 167–192
9. Engel, J. & Prockop, D.J. (1991) The zipper-like folding of collagen triple helices and the effects of mutations that disrupt the zipper. *Annu. Rev. Biophys. Biophys. Chem.* **20**, 137–152
10. Mould, A.P., Hulmes, D.J.S., Holmes, D.F., Cummings, C., Sear, C.H.J. & Chapman, J.A. (1990) D-periodic assemblies of type I procollagen. *J. Mol. Biol.* **211**, 581–594
11. Seyedin, S.M. & Rosen, D.M. (1990) Matrix proteins of the skeleton. *Curr. Opin. Cell Biol.* **2**, 914–919
12. Hojima, Y., McKenzie, J., van der Rest, M. & Prockop, D.J. (1989) Type I N-proteinase from chick embryo tendons. *J. Biol. Chem.* **264**, 11336–11345
13. Kessler, E., Mould, A.P. & Hulmes, D.J.S. (1990) Procollagen type I C-proteinase enhancer is a naturally occurring connective tissue glycoprotein. *Biochem. Biophys. Res. Commun.* **173**, 81–86
14. Kagan, H.M. & Trackman, P.C. (1991) Properties and function of lysyl oxidase. *Am. J. Resp. Cell Mol. Biol.* **5**, 206–210
15. Torre-Blanco, A., Adachi, E., Hojima, Y., Wootton, J.A.M., Minor, R.R. & Prockop, D.J. (1992) Temperature-induced post-translational over-modification of type I procollagen.

Effects of overmodification of the protein on the rate of cleavage by procollagen N-proteinase and on self-assembly of collagen into fibrils. *J. Biol. Chem.* **267**, 2650–2655

16. Kadler, K.E., Hojima, Y. & Prockop, D.J. (1990) Collagen fibrils *in vitro* grow from pointed tips in the C- to N-terminal direction. *Biochem. J.* **268**, 339–343

17. Na, G.C. (1989) Monomer and oligomer of type I collagen: molecular properties and fibril assembly. *Biochemistry* **28**, 7161–7167

18. Oosawa, F. & Asakura, S. (1975) *Thermodynamics of the Polymerization of Protein*, pp. 1–204, Academic Press, London

19. Birk, D.E., Zycband, E.I., Winkelmann, D.A. & Trelstad, R.L. (1989) Collagen fibrillogenesis *in situ* : fibrils segments are intermediates in matrix assembly. *Proc. Natl. Acad. Sci. U.S.A.* **86**, 4549–4553

20. Jones, E.Y. & Miller, A. (1991) Analysis of structural design features in collagen. *J. Mol. Biol.* **218**, 209–219

21. Hulmes, D.J.S., Holmes, D.F. & Cummings, C. (1985) Crystalline regions in collagen fibrils. *J. Mol. Biol.* **184**, 473–477

22. Birk, D.E., Fitch, J.M., Babiarz, J., Doand, J. & Linsenmayer, T.F. (1990) Collagen fibrillogenesis *in vitro*: interactions of type I and V collagen regulate fibril diameter. *J. Cell Sci.* **95**, 649–657

23. Scott, J.E. (1988) Proteoglycan–fibrillar collagen interactions. *Biochem. J.* **252**, 313–323

24. Fleischmajer, R., Fisher, L.W., MacDonald, E.D., Jacobs, L., Perlish, J.S. & Termine, J.D. (1991) Decorin interacts with fibrillar collagen of embryonic and adult human skin. *J. Struct. Biol.* **106**, 82–90

25. Hulmes, D.J.S., Kadler, K.E., Mould, A.P., Hojima, Y., Holmes, D.F., Cummings, C., Chapman, J.A. & Prockop, D.J. (1989) Pleomorphism in type I collagen fibrils produced by persistence of the procollagen N-propeptide. *J. Mol. Biol.* **210**, 337–345

26. Kadler, K.E., Torre-Blanco, A., Adachi, E., Vogel, B.E., Hojima, Y. & Prockop, D.J. (1991) A type I collagen with substitution of a cysteine for glycine-748 in the alpha-1(I) chain co-polymerizes with normal type-I collagen and can generate fractal-like structures. *Biochemistry* **30**, 5081–5088

27. Yurchenco, P.D. & Schittny, J.C. (1990) Molecular architecture of basement membranes. *FASEB J.* **4**, 1577–1590

28. Parente, M.G., Chung, L.C., Ryynanen, J, Woodley, D.T., Wynn, K.C., Bauer, E.A., Mattei, M.-G., Chu, M.-L. & Uitto, J. (1991) Human type VII collagen: cDNA cloning and chromosomal mapping of the gene. *Proc. Natl. Acad. Sci. U.S.A.* **88**, 6931–6935

29. Blumberg, B., Fessler, L.I., Kurkinen, M. & Fessler, J.H. (1986) Biosynthesis and supramolecular assembly of procollagen IV in neonatal lung. *J. Cell Biol.* **103**, 1711–1719

30. Kwan, A.P.L., Cummings, C.E., Chapman, J.A. & Grant, M.E. (1991) Macromolecular organization of chicken type X collagen *in vitro*. *J. Cell Biol.* **114**, 597–604

31. Shaw, L.M. & Olsen, B.R. (1991) FACIT collagens: diverse molecular bridges in extracellular matrices. *Trends Biochem. Sci.* **16**, 191–194

32. Ayad, S., Marriott, A., Brierley, V.H. & Grant, M.E. (1991) Mammalian cartilage synthesizes both proteoglycan and non-proteoglycan forms of type IX collagen. *Biochem. J.* **278**, 441–445

33. Yamagata, M. Yamada, K.M., Yamada, S.S., Shinomura, T., Tanaka, H., Nishida, Y., Obara, M. & Kimata, K. (1991) The complete structure of type XII collagen shows a chimeric molecule with reiterated fibronectin type III motifs, von Willebrand factor A motifs, a domain homologous to a noncollagenous region of type IX collagen, and short collagenous domains with an Arg-Gly-Asp site. *J. Cell Biol.* **115**, 209–221

34. Gordon, M.K., Castagnola, P., Dublet, B., Linsenmayer, T.F., van der Rest, M., Mayne, R. & Olsen, B.R. (1991) Cloning of a cDNA for a new member of the class of fibril-associated collagens with interrupted triple helices. *Eur. J. Biochem.* **201**, 333–338

35. Tikka, L., Elomaa, O., Pihlajaniemi, T. & Tryggvason, K. (1991) Human α1(XIII) gene. *J. Biol. Chem.* **266**, 17713–17719

36. Colombatti, A. & Bonaldo, P. (1991) The superfamily of proteins with von Willebrand factor type A-like domains: one theme common to components of extracellular matrix, hemostasis, cellular adhesion, and defense mechansims. *Blood* **77**, 2305–2315

37. Stokes, D.G., Saitta, B., Timpl, R. & Chu, M.-L. (1991) Human α3(VI) collagen gene. *J. Biol. Chem.* **266**, 8626–8633

38. Doliana, R., Bonaldo, P. & Colombatti, A. (1991) Multiple forms of chicken α3(VI) collagen chain generated by alternative splicing in type A repeated domains. *J. Cell Biol.* **111**, 2197–2205

39. Sekiguchi, K., Maeda, T. & Titani, K. (1991) Artificial cell adhesive proteins. *Essays Biochem.* **26**, 39–48

40. Sellar, G.C., Blake, D.J. & Reid, K.B.M. (1991) Characterization and organization of the genes encoding the A-, B- and C-chains of human complement subcomponent C1q. *Biochem. J.* **274**, 481–490

41. Thiel, S. & Reid, K.B.M. (1989) Structures and functions associated with the group of mammalian lectins containing collagen-like sequences. *FEBS Lett.* **250**, 78–84

42. Krejci, E., Coussen, F. Duval, N., Chatel, J.-M., Legay, C., Puype, M., Vandekerckhove, J., Cartaud, J., Bon, S. & Massoulié, J. (1991) Primary structure of a collagen tail peptide of *Torpedo* acetylcholinesterase: co-expression with catalytic subunit induces the production of collagen-tailed forms in transfected cells. *EMBO J.* **10**, 1285–1293

43. Kodama, T., Freeman, M., Rohrer, L., Zabrecky, J., Matsudaira, P. & Krieger, M. (1990) Type I macrophage scavenger receptor contains α-helical and collagen-like coiled coils. *Nature (London)* **343**, 531–535

44. Spissinger, T., Schafer, K.P. & Voss, T. (1991) Assembly of the surfactant protein SP-A. *Eur. J. Biochem.* **199**, 65–71

45. Raspanti, M., Ottani, V. & Ruggeri, A. (1989) Different architectures of the collagen fibril: morphological aspects and functional implications. *Int. J. Biol. Macromol.* **11**, 367–371

46. Bruns, R. R. (1984) Beaded filaments and long-spacing fibrils: relation to type VI collagen. *J. Ultrastruct. Res.* **89**, 136–145

47. Vaughan, L., Mendler, M., Huber, S., Bruckner, P., Winterhalter, K.H., Irwin, M.I. & Mayne, R. (1988) D-periodic distribution of collagen type IX along cartilage fibrils. *J. Cell Biol.* **106**, 991–997

5

The repair of DNA alkylation damage by methyltransferases and glycosylases

Leona D. Samson

Department of Molecular and Cellular Toxicology, Harvard School of Public Health, Boston, MA 02115, U.S.A.

DNA ALKYLATION DAMAGE

The genome of every organism continually sustains DNA damage which, if left unrepaired, contributes to cell death, mutation, chromosome damage, ageing and carcinogenesis. Consequently, numerous DNA repair pathways have evolved to prevent the accumulation of DNA damage in the genome[1], and these pathways appear to be highly conserved among bacteria, yeast, insect, fish and mammalian cells. Alkylating agents represent one the most abundant classes of DNA-damaging agents in our environment and are potent mutagens and carcinogens. These agents occur widely, for example, in our food, in fuel combustion products, tobacco smoke, and some cosmetic products[2]; ironically, certain alkylating agents are also used for the chemotherapeutic treatment of cancer patients. The characterization of DNA damage produced by methylating and ethylating agents indicates that methyl and ethyl groups can be transferred to almost every oxygen and nitrogen in DNA, producing a dozen different adducts[3] (Figure 1a). Judging from their locations on the DNA molecule, some adducts protrude into the major groove of the double helix (O^6-methylguanine, O^4-methylthymine, 7-methyladenine and 7-methylguanine), some protrude into the

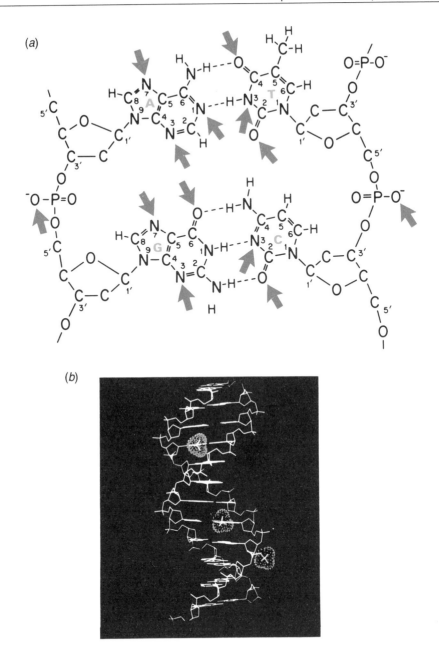

Figure 1. DNA alkylation

(a)The arrows indicate all possible sites of DNA alkylation. From left to right, top to bottom, the adducts produced by methylation at these sites are 7-methyladenine, O^4-methylthymine, 3-methyladenine, 1-methyladenine, 3-methylthymine, O^2-methylthymine, methylphosphotriester, 7-methylguanine, O^6-methylguanine,methylphosphotriester, 3-methylguanine, 3-methylcytosine, and O^2-methylcytosine. (b) The van der Waals' radii of three methyl groups on the DNA double helix, in the minor groove (O^2-methylthymine, top), the major groove (O^6-methylguanine, middle), and on the sugar-phosphate backbone (methylphosphotriester, bottom).

minor groove (O^2-methylcytosine, O^2-methylthymine, 3-methyladenine and 3-methyl-guanine), and methylphosphotriester adducts protrude from the sugar phosphate backbone; a methyl adduct in each of these three areas is illustrated in Figure 1*b*. The ease with which each of the twelve alkylated DNA adducts is produced varies widely and depends in part upon the nucleophilicity of the relevant oxygen or nitrogen and upon the particular alkylating agent. Generally speaking, for agents like *N*-methyl-nitrosourea and *N*-methyl-*N'*-nitro-*N*-nitrosoguanidine, 7-methylguanine is produced at the highest levels (60–80%), O^6-methylguanine, methylphosphotriesters and the 3-methylpurines at intermediate levels (5–12%) and the remaining adducts at very low levels (up to 2%)[3,7]. It will become evident below that the toxic effects of alkylating agents are caused primarily by some of the intermediate- and low-abundance DNA adducts.

Every type of organism so far tested has been found to possess efficient DNA repair mechanisms to ensure that particular alkylated oxygens and nitrogens do not accumulate in the genome. Alkylation-specific repair is executed by methyltransferases which repair O^6-methylguanine, O^4-methylthymine and methylphosphotriester DNA adducts, and glycosylases which catalyse the release of a variety of *N*-alkylated and O^2-alkylated bases from DNA. These methyltransferases and glycosylases, and their biological functions, have been best characterized in *Escherichia coli*, but their counter-parts have been identified in eukaryotes, most notably in humans. This essay details our current understanding of alkylation-specific DNA repair and its biological con-sequences in *E. coli*, and summarizes our current understanding of DNA alkylation repair in mammalian cells.

DNA REPAIR METHYLTRANSFERASES

The inducible Ada methyltransferase of E. coli

The *E. coli* Ada protein was the first DNA repair methyltransferase to be discovered, and is the one best characterized. The name Ada derives from the fact that it plays a central role in the **Ada**ptive response of *E. coli* to alkylating agents. *E. coli* grown in low, non-toxic levels of methylating agents adapt to acquire tremendous resistance to the biological effects of alkylating agents[4], and this adaptive resistance is achieved by inducing the expression of at least four genes, among them the *ada* methyltrans-ferase gene itself[5,6]. Early studies showed that alkylation-adapted *E. coli* rapidly removes O^6-methylguanine from its genome, and the kinetics of O^6-methylguanine removal were correctly interpreted to indicate that the protein that carried out O^6-methylguanine repair was efficient and could act only once. The protein responsible for O^6-methylguanine repair was subsequently identified as a DNA repair methyl-transferase and shown to repair *O*-alkyl adducts via a suicide mechanism[5]. Methyl-transferases transfer methyl groups from particular methylated oxygens in DNA to an active site cysteine residue to form *S*-methylcysteine (Figure 2), and it appears that this methyl transfer is irreversible; thus, strictly speaking, DNA repair methyl-transferases are not enzymes because they are consumed in the reaction. The Ada methyltransferase is a 354-amino-acid, 39 kDa protein and has two active sites centred on Cys-69 and Cys-321; the roles played by each active site cysteine are described below.

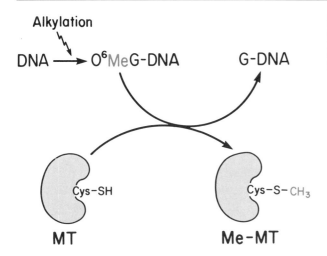

Alkylation

DNA $\longrightarrow$ O^6MeG-DNA   G-DNA

Cys-SH   Cys-S-CH_3

MT   Me-MT

Figure 2. Transfer of a methyl group from alkylated DNA to an O^6-methylguanine DNA repair methyltransferase

Cys-321 transfers a methyl group from either O^6-methylguanine or O^4-methylthymine DNA lesions, which are formed at a ratio of roughly 10:1 by most methylating agents[3,7]. Since O^6-methylguanine and O^4-methylthymine have a high probability of mispairing during replication (with thymine and guanine respectively) their repair by the Ada methyltransferase prevents the induction of G:C to A:T and A:T to G:C transition mutations[8,9] (Figure 3). Note that the methyl adducts on O^6-methylguanine and O^4-methylthymine both lie in the major groove of the DNA double helix and that the Ada protein is believed to track along double-stranded DNA with its C-terminal end in the major groove; confirmation of this hypothesis awaits X-ray crystallographic and n.m.r. analyses of the Ada methyltransferase bound to its substrates.

Cys-69 transfers a methyl group from methylphosphotriester adducts[5]. Methylphosphotriesters are not thought to cause cell death or mutation in E. coli but their repair by the Ada methyltransferase has profound biological consequences. Methylation at Cys-69 is believed to cause a conformational change in the Ada protein because it vastly increases its ability to bind to the AAAGCGCA DNA sequence located just upstream of the ada-alkB operon (Figure 3); upon binding to this so-called "Ada box" sequence the Cys-69-methylated Ada protein stimulates ada-alkB transcription by facilitating the binding of RNA polymerase to the ada-alkB promoter[5,6]. (The product of the alkB gene, located downstream from ada, protects E. coli against killing by certain alkylating agents, but the mechanism is unknown.) The Ada methyltransferase probably binds similar sequences upstream of the alkA gene and the aidB gene because they too are induced in alkylation-adapted E. coli and their induction depends upon the presence of a functional Ada protein[5,6]. The function of the alkA gene product is described below, and the function of the AidB protein is not known (Figure 3).

In summary, the E. coli Ada methyltransferase protein prevents transition mutations by repairing O^6-methylguanine and O^4-methylthymine, acts as a sensor for DNA alkylation damage by repairing methylphosphotriesters, and activates the transcription of at least four genes whose products enable E. coli to recover from the toxic effects of alkylating agents (Figure 3).

The constitutive Ogt methyltransferase of E. coli

In the non-adapted state, *E. coli* contains one or two Ada methyltransferase molecules per cell. It takes roughly 1 hour of growth in sublethal levels of alkylating agents to induce *ada* expression fully (producing about 3000 molecules per cell) during which time cells replicate their genome at least once[4-6]. The time taken to induce *ada* is therefore a potentially very vulnerable period during which unrepaired O^6-methylguanine and O^4-methylthymine lesions could produce transition mutations[8,9]. It is therefore not surprising (in retrospect) that *E. coli* constitutively expresses another DNA methyltransferase, and this methyltransferase provides instant protection against mutagenesis by very low levels of DNA alkylation damage[10]. This second methyltransferase is present at about 30 molecules per cell and is encoded by the *ogt* gene (Figure 3)[11,12]. The Ogt methyltransferase is roughly half the size of the Ada protein and the amino acid sequence of its C-terminal half, containing the probable methyl acceptor (Cys-139), is very similar to that of the roughly 100 amino acids surrounding the Cys-321 active site of Ada, but is not similar to the region surrounding the Cys-69 active site residue of Ada[11].

As one would predict from the similarity of the Ogt active site region to the Cys-321 region of Ada (and its lack of similarity to the Cys-69 region), Ogt can transfer methyl groups from either O^6-methylguanine or O^4-methylthymine DNA lesions, but cannot transfer methyl groups from methylphosphotriester lesions. However, the relative affinities of Ada and Ogt for O^6-methylguanine versus O^4-methylthymine show an interesting difference. The affinity of Ada Cys-321 and Ogt Cys-139 for O^6-methylguanine is roughly similar; however, while Ogt has about a 3-fold higher affinity for

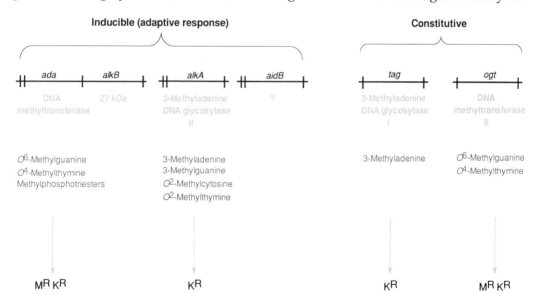

Figure 3. The genes and proteins known to be involved in the specific repair of DNA alkylation damage in *Escherichia coli*

Where possible, the alkylated DNA substrates, and the biological consequences of repairing those substrates, are indicated; M^R and K^R indicate mutation resistance and killing resistance respectively.

O^4-methylthymine than O^6-methylguanine, Ada has a 22-fold lower affinity for O^4-methylthymine than O^6-methylguanine, as judged from their ability to act on these altered bases *in vitro*[13]. That Ogt can repair both O^4-methylthymine and O^6-methylguanine efficiently may allow the repair of both adducts in unadapted *E. coli*; if Ogt were like Ada, with a much higher preference for O^6-methylguanine than O^4-methylthymine, the O^4-methylthymine adduct (normally present at one-tenth the level of O^6-methylguanine) might seldom be repaired due to the continuous inactivation of Ogt by O^6-methylguanine. Transcription of *ogt* is not greatly induced in adapted *E. coli* and adaptation actually depletes the cell of active Ogt molecules because of suicide inactivation[10]. However, adapted *E. coli* cells contain several thousand Ada molecules and the excess Ada methyltransferase presumably compensates for its low O^4-methylthymine affinity, so that both O^6-methylguanine and O^4-methylthymine can be repaired in adapted *E. coli*.

The biological roles of the E. coli DNA repair methyltransferases

The biological roles of Ada and Ogt were deduced by characterizing the phenotypes of *E. coli* strains bearing *ada* and *ogt* mutations. These studies clearly demonstrate that Ada and Ogt protect cells from both the mutagenic and the cytotoxic effects of alkylating agents[5,12] (Figure 3). Unrepaired O^6-methylguanine and O^4-methylthymine adducts frequently direct DNA polymerase to incorporate thymine and guanine (respectively) during replication. The rapid removal of O^6-methylguanine and O^4-methylthymine by methyltransferases prevents the transition mutations that would result from misincorporated thymine and guanine; thus, Ogt prevents mutation at low levels of DNA alkylation damage (produced by endogenous and exogenous alkylating agents) and Ada is recruited to prevent mutation at higher levels of alkylation damage[12]. *E. coli* that do not express any Ada and Ogt methyltransferase activity suffer an elevated spontaneous mutation rate. Mutants arise in stationary phase populations of *ada⁻ ogt⁻ E. coli* at a higher rate than in wild type *E. coli*, and the extra mutants include G:C to A:T and A:T to G:C transitions which are presumably caused by O^6-methylguanine or O^4-methylthymine produced by endogenous metabolites acting as alkylating agents[12]. It has been suggested (and it is generally believed) that metabolites such as *S*-adenosylmethionine provide an endogenous and continuous source of DNA alkylation damage; the elevated rate of spontaneous mutation in methyltransferase-deficient cells constitutes the first *in vivo* evidence to support this idea.

How methyltransferases prevent alkylation-induced cytotoxicity is not fully understood, and conventionally O^6-methylguanine and O^4-methylthymine have been considered mutagenic but not lethal. Clearly the Ada methyltransferase indirectly prevents alkylation-induced cell killing by switching on transcription of the *alkA* DNA glycosylase gene whose product repairs the cytotoxic 3-methyladenine DNA lesion (see below). However, the O^6-methylguanine/O^4-methylthymine Ogt methyltransferase, which is not thought to regulate the expression of other genes, also provides resistance to the killing effects of alkylating agents[12]. How can we explain the observation that O^6-methylguanine and O^4-methylthymine appear to cause both killing and mutation, since a dead cell cannot go on to form a mutant cell? There are several possibilities. Cell death might simply be due to a fraction of the alkylation-induced point mutations being lethal mutations. However, some simple calculations indicate that in order for this to be true a substantial fraction of these point mutations would have to be lethal, and this is probably not true. Alternatively, cell death might result from a fraction of the O-alkyl lesions being located in critical regions of the genome; for example, if O^6-methylguanine, O^4-methylthymine or methylphosphotriesters in DNA replication origins could not be repaired efficiently and the lesion(s) interfered with initiation of DNA replication, then O-alkylation in these "critical" regions would cause cell death[14]. Finally, it is difficult to discount the possibility that the Ogt protein may switch on the transcription of hitherto unidentified genes whose products alleviate the cytotoxic effects of alkylating agents.

Mammalian DNA repair methyltransferases

Mammalian cells appear to express a single DNA repair methyltransferase which has been called MGMT[15]. The human *MGMT* cDNA was recently cloned, the *MGMT* gene mapped to human chromosome 10, and the MGMT protein purified to homogeneity. Comparison of MGMT with the *E. coli* methyltransferases reveals both similarities and differences. The MGMT active site region (around Cys-145) shows amino sequence similarity to the O^6-methylguanine/O^4-methylthymine active site regions of both Ada and Ogt. As this sequence similarity predicts, MGMT does not repair methylphosphotriester lesions but does repair O^6-methylguanine lesions. However, whether MGMT repairs O^4-methylthymine DNA lesions remains controversial; three groups report that MGMT recognizes O^4-methylthymine in DNA, and several others report that it does not[13,16]. MGMT probably has an extremely low affinity for O^4-methylthymine which is detectable under certain conditions *in vitro* but which may have no relevance *in vivo*. Several MGMT-deficient human cell lines have been identified and characterization of their phenotype indicates that O^6-methylguanine in the human genome causes cell death, mutation and chromosome damage[7]; as in *E. coli*, the exact mechanism by which O^6-methylguanine causes cell death is not yet understood. In certain cell types *MGMT* gene expression is induced in response to DNA damage[17]. However, this induction differs from that of *ada* in that *MGMT* protein and mRNA levels can only be increased up to 5-fold (versus several-hundred-fold for *ada*), and *MGMT* is induced in response to many different types of DNA damage, whereas *ada* induction is alkylation-specific. Thus, the molecular mechanism of methyltransferase induction in mammalian cells is likely to be different from that in *E. coli*.

DNA GLYCOSYLASES

DNA glycosylases catalyse the hydrolytic cleavage of the *N*-glycosylic bond linking an abnormal or damaged base to deoxyribose in the sugar-phosphate DNA backbone (Figures 1 and 4), and upon hydrolysis produce an apyrimidinic or apurinic (AP) site in DNA. The original undamaged DNA sequence can then be restored by the consecutive action of AP endonuclease, exonuclease, DNA polymerase and DNA ligase enzymes; the collective action of these five types of enzyme is called base excision DNA repair[1] and is illustrated in Figure 4. At least eight types of DNA glycosylase have been identified, each of which is specific for the removal of one or more abnormal DNA bases. Below are described two types of *E. coli* glycosylase and one mammalian glycosylase that specifically excise alkylated bases from DNA.

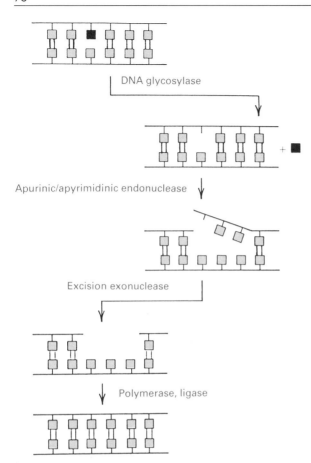

Figure 4. Base excision DNA repair

The AlkA and Tag 3-methyladenine DNA glycosylases of E. coli

Analogous to the Ogt and Ada methyltransferases, *E. coli* expresses one constitutive and one alkylation-inducible DNA glycosylase for the repair of alkylated DNA bases[5,6] (Figure 3). The constitutively expressed Tag glycosylase was the first DNA alkylation repair enzyme to be discovered. Tag has a narrow substrate specificity, excising only 3-methyladenine from alkylated DNA. The alkylation-inducible AlkA glycosylase has a somewhat broader substrate specificity, efficiently excising 3-methyladenine, 3-methylguanine, O^2-methylcytosine and O^2-methylthymine and inefficiently excising 7-methylpurines from alkylated DNA. The methyl groups on each of the alkylated bases recognized by Tag and AlkA protrude into the minor groove of the DNA double helix, and it is presumed that these glycosylases track along the minor groove in search of their substrates. Surprisingly, Tag and AlkA show no amino acid sequence similarities despite the fact that they both catalyse the release of 3-methyladenine from DNA; their enzymic mechanisms are still unknown.

The biological roles of the E. coli DNA glycosylases

The biological roles of AlkA and Tag were deduced by characterizing the phenotypes of *E. coli* strains bearing *alkA* and *tag* mutations. *E. coli* mutants deficient in either the Tag or the AlkA glycosylase activities become extremely sensitive to the killing effects (but not the mutagenic effects) of alkylating agents[5,6]. The mechanism by which 3-methyladenine causes cell death in *E. coli* appears to be by the inhibition of DNA replication; unlike the O^6-methylguanine and O^4-methylthymine lesions, 3-methyladenine presents a block to DNA polymerases and thus to cell division. (Non-dividing *E. coli* are scored as dead because they do not form a visible colony on agar.) The killing sensitivity of AlkA glycosylase-deficient *E. coli* mutants is almost completely suppressed by the over-expression of the Tag glycosylase. This indicates that 3-methyladenine is the major toxic lesion produced by methylating agents, and that the other lesions repaired efficiently by AlkA (namely 3-methylguanine and O^2-methylpyrimidines) are relatively minor toxic lesions. It is not yet known whether the low abundance of these lesions (each less than 1%), or their inability to block DNA replication, is responsible for their minor contribution to alkylation-induced cell death.

Mammalian 3-methyladenine DNA glycosylases

3-Methyladenine DNA glycosylases have been partially purified from several mammalian cell types and a 27 kDa calf thymus glycosylase was purified to homogeneity[18]. More recently, cDNAs encoding rat, human and mouse 3-methyladenine DNA glycosylases were cloned by their expression in *E. coli* and by their ability to rescue *alkA/tag* glycosylase deficient *E. coli* mutants from the toxic effects of alkylating agents[19,20]; these three mammalian glycosylases share 70–90% amino acid sequence identity, but show no significant similarities to the *E. coli* AlkA and Tag glycosylases. The results of substrate specificity studies vary between laboratories. By definition, the mammalian 3-methyladenine DNA glycosylases repair 3-methyladenine, and several labs have shown that they also release 3-methylguanine and 7-methylguanine from alkylated DNA; the O^2-methylpyrimidines have not yet been examined. The biological role of the 3-methyladenine DNA glycosylase in mammalian cells is presumed to be the same as that in *E. coli*, i.e., to protect cells against the cytotoxic effects of alkylating agents. Confirmation of this presumption awaits the generation of mutant mammalian cell lines deficient in 3-methyladenine DNA glycosylase activity, and the recently cloned glycosylase now makes this feasible.

I thank the members of my laboratory for their critical comments. Because of the limit on the number of references in this article, I must apologize to any colleagues who feel their papers should have been cited; I have tried to cite reviews and recent papers that should direct the reader to all the appropriate citations. My work is supported by an American Cancer Society Faculty Research Award and studies from my laboratory are supported by National Institutes of Health Grant CA55042 and National Institute of Environmental Health Science Grant ES03926.

REFERENCES

1. Friedberg, E.F. (1985) *DNA Repair*, Freeman, San Francisco
2. Hoffman, D. & Hecht, S. (1985) Nicotine-derived N-nitrosamines and tobacco-related cancer: current status and future directions, *Cancer Res.* **45**, 935–934
3. Montesano, R. (1981) Alkylation of DNA and tissue specificity in nitrosamine carcinogenesis. *J. Supramol. Struct. Cell. Biochem.* **17**, 259–273
4. Samson, L. & Cairns, J. (1977) A new pathway for DNA repair in *Escherichia coli*. *Nature (London)* **267**, 281–283

5. Lindahl, T., Sedgwick, B., Sekiguchi, M. & Nakabeppu, Y. (1988) Regulation and expression of the adaptive response to alkylating agents. *Annu. Rev. Biochem.* **57**, 133–157

6. Shevell, D.E., Friedman, B.M. & Walker, G.C. (1990) Resistance to alkylation damage in *Escherichia coli*: role of the Ada protein in induction of the adaptive response. *Mutation Res.* **233**, 57–72

7. Day, R.S., Babich, M.A., Yarosh, D.B. & Scudiero, D.A. (1987) The role of O^6-methylguanine in human cell killing, sister chromatid exchange induction and mutagenesis: a review. *J. Cell Sci. Suppl.* **6**, 333–353

8. Loechler, E.L., Green, C.L. & Essigmann, J.M. (1984) *In vivo* mutagenesis by O^6-methylguanine built in a unique site in a viral genome. *Proc. Natl. Acad. Sci. U.S.A.* **81**, 6271–6275

9. Preston, B.D., Singer, B. & Loeb, L.A. (1986) Mutagenic potential of O^4-methylthymine *in vivo* determined by an enzymatic approach to site specific mutagenesis. *Proc. Natl. Acad. Sci. U.S.A.* **83**, 8501–8505

10. Rebeck, G.W., Coons, S., Carroll, P. & Samson, L. (1988) A second DNA methyl transferase in *Escherichia coli*. *Proc. Natl. Acad. Sci. U.S.A.* **85**, 3039–3043

11. Potter, P.M., Wilkinson, M.C., Fitton, J., Carr, F.J., Brennand, J., Cooper, D.P. & Margison, G.P. (1987) Characterization and nucleotide sequence of *ogt*, the O^6-alkylguanine-DNA alkyltransferase gene. *Nucleic Acids Res.* **15**, 9177–9193

12. Rebeck, G.W. & Samson, L. (1991) Increased spontaneous mutation and alkylation sensitivity in *Escherichia coli* strains lacking the Ogt O^6-methyltransferase. *J. Bacteriol.* **173**, 2068–2076

13. Sassanfar, M., Dosanjh, M.K., Essigmann, J.M. & Samson, L. (1991) Relative efficiencies of the bacterial, yeast and human DNA methyltransferases for the repair of O^6-methylguanine and O^4-methylthymine. *J. Biol. Chem.* **266**, 2767–2771

14. Bignami, M. & Lane, D.P. (1990) O^6-Methylguanine in the SV40 origin of replication inhibits binding but increases unwinding by viral large T antigen. *Nucleic Acids Res.* **9**, 3089–3103

15. Pegg, A.E. (1990) Mammalian O^6-alkylguanine-DNA alkyltransferase: regulation and importance in response to alkylating carcinogenic and therapeutic agents. *Cancer Res.* **50**, 6119–6129

16. Brent, T.P., Dolan, M.E., Fraenkel-Conrat, H., Hall, J., Karran, P., Laval, F., Margison, G.P., Montesano, R., Pegg, A.E., Potter, P.M., Singer, B.,Swenberg, J.A. & Yarosh, D.B. (1988) Repair of O^6-alkylpyrimidines in mammalian cells: a present consensus. *Proc. Natl. Acad. Sci. U.S.A.* **85**, 1759–1762

17. Fritz, G., Tano, K., Mitra, S. & Kaina, B. (1991) Inducibility of the DNA repair gene encoding O^6-methylguanine-DNA methyltransferase in mammalian cells by DNA-damaging treatments. *Mol. Cell. Biol.* **11**, 4660–4668

18. Male, R., Haukanes, B.I., Helland, D.E. & Kleppe (1987) Substrate specificity of 3-methyladenine-DNA glycosylase from calf thymus. *Eur. J. Biochem.* **165**, 13–19

19. O'Connor, T.R. & Laval, F. (1990) Isolation and structure of a cDNA expressing a mammalian 3-methyladenine-DNA glycosylase. *EMBO J.* **9**, 3337–3342

20. Samson, L., Derfler, B., Boosalis, M. & Call, K. (1991) Cloning and characterization of a 3-methyladenine DNA glycosylase cDNA from human cells whose gene maps to chromosome 16. *Proc. Natl. Acad. Sci. U.S.A.* **88**, 9127–9131

6

Endothelins

Tomoh Masaki and Masashi Yanagisawa

Department of Pharmacology, Faculty of Medicine, Kyoto University, Kyoto 606, Japan

INTRODUCTION

The major function of the blood vessels is to supply blood to peripheral tissues. To this end, blood vessels control the blood stream through constriction and dilation. It has long been thought that this function of the blood vessel is regulated by the vasoconstrictor nerves of the autonomic nervous system terminating in the smooth muscle layer and that the endothelium, which lines the luminal side of the vascular wall, was merely a simple barrier preventing direct contact between blood and the tissue of the vascular wall.

However, the discovery of endothelium-derived relaxing factor (EDRF) (now identified with nitric oxide) by Furchgott & Zawadki[1] led to the new concept that the endothelium also plays an important role in the regulation of vasoconstriction and vasodilation[2]. In order to do this, the endothelium produces prostacyclin and EDRF as vasodilators, and thromboxane A_2 and endothelin as vasoconstrictors[2].

Endothelin was initially found in the conditioned medium of cultured endothelial cells[3]. It is the most potent vasoconstrictive peptide known so far. Since administration of endothelin into mammalian animals elicits a sustained increase in blood pressure, endothelin is thought to play an important role in vasospasm and the maintainance of blood pressure.

STRUCTURE OF ENDOTHELIN

Endothelin consists of 21 amino acid residues with free N- and C-termini. There are four cysteine residues, forming two intramolecular disulphide bonds between amino acids 1 and 15 and 3 and 11. The successive hydrophobic amino acid residues at the C-terminus are important in exerting vasoconstriction.

Analysis of the human genomic gene for endothelin revealed the existence of three

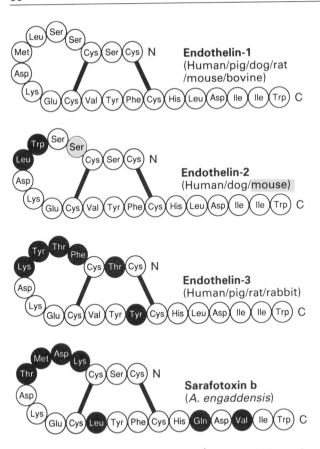

Figure 1. Structures of endothelins and sarafotoxin S6b
In mouse endothelin-2, Ser-4 of endothelin-1 is replaced by Asn-4.

distinct isotypes of endothelin[4]. One of these has the same sequence as the native endothelin initially isolated from the conditioned medium of cultured endothelial cells, and is designated endothelin-1. The other two were designated endothelin-2 and endothelin-3 (Figure 1). Interspecies differences among the three endothelins have not been found to any great extent, except in the case of mouse endothelin-2. The three distinct isopeptides show slightly different biological activities. Endothelin-2 and endothelin-1 are more potent than endothelin-3 in exerting vasoconstriction and pressor response, whereas endothelin-3 is relatively potent in vasodilation activity. Northern blot analysis confirmed that the mRNAs of the three distinct endothelins were expressed in different proportions not only in the vascular system but also in non-vascular tissues[5].

Interestingly, the amino acid sequences of a family of rare snake venoms, the sarafotoxins, are very similar to the sequences of the endothelins[6] (Figure 1).

PRODUCTION OF ENDOTHELIN

Endothelin is produced not only in endothelial cells but also in non-vascular tissues, including lung, intestine, kidney, adrenal gland, pancreas, spleen, heart, eye, placenta and central nervous tissue[7]. The distribution of the endothelin peptide is very similar to that of the endothelin receptor[8], suggesting that endothelin is a local hormone.

The three isotypes of endothelins are produced in various tissues in different proportions. Endothelial cells predominantly produce endothelin-1.

In the central nervous system, endothelin-1 is produced in the cortex, cerebellum, hippocampus, hypothalamus, amygdala, medulla oblongata, spinal cord and dorsal root ganglion[7,9]. The distribution is very similar to that of substance P, β-preprotachykinin, calcitonin-gene-related peptide and cholecystokinin. As endothelin-1 exists in the intermediolateral cell column of human spinal cord where preganglionic fibres of sympathetic neurons emerge[9], endothelin-1 may be a neuropeptide modulator of cardiovascular function. Endothelin may also play an important role in the pituitary gland, in releasing gonadotropins or growth hormone or in water balance[10,11].

Preproendothelin-1 mRNA was induced in endothelial cells by thrombin, transforming growth factor β, interleukin-1, tumour necrotizing factor α, angiotensin II, vasopressin and by the shear stress of the blood stream[7]. Production of endothelin-1 in endothelial cells is mediated by an increase in intracellular free calcium ion induced by these factors. Since phorbol ester induces production of endothelin-1, protein kinase C activation may also be involved in this mechanism. On the other hand, an increase in the cyclic GMP level elicited by nitric oxide or atrial natriuretic peptide inhibits the thrombin-induced production of endothelin-1[7].

The half-life of endothelin-1 mRNA is short, similar to that of the cytokines. However, the lifespan of the mRNA was significantly lengthened in the presence of cycloheximide, suggesting the existence of an inhibitory protein that destabilizes endothelin-1 mRNA.

Figure 2. Biosynthetic pathway of human endothelin-1

38-Amino-acid human big endothelin-1 is produced from proendothelin-1 and cleaved into endothelin-1 by endothelin-converting enzyme. Endothelin-1 converting enzyme is specific for the cleavage site Trp–Val; big endothelin-2 and big endothelin-3 are probably cleaved by a different converting enzyme(s).

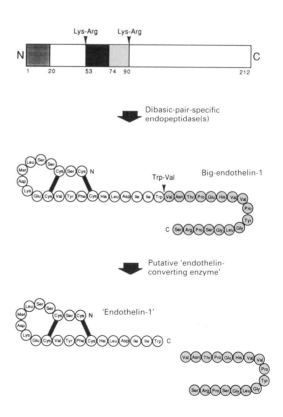

The precursor of the peptide is synthesized in endothelial cell cytoplasm. Since the signal peptide sequence is at the N-terminus, proendothelin-1 must cross the inner membrane and be stored temporarily in vesicles. In porcine endothelium, proendo-thelin-1 is initially cleaved in a vesicle by an endopeptidase specific for a pair of dibasic amino acids to produce an intermediate 39-amino-acid peptide, designated big endothelin-1, which is in turn cleaved into mature endothelin-1 by a putative endothelin-converting enzyme[3] (Figure 2). Human big endothelin-1 is 38 amino acid residues long. Both big endothelin-1 and endothelin-1 are secreted into the extracellular space. Since there are no secretory granules in endothelial cells, endothelin-1 and big endothelin-1 must be secreted through a consecutive pathway.

Although the endothelin-converting enzyme has so far not been isolated and puri-fied, the enzyme is estimated to be a membrane-bound metalloproteinase of molecular mass about 100 kDa. Its activity is inhibited by EDTA, o-phenanthroline and phos-phoramidon[12].

STRUCTURE OF THE ENDOTHELIN RECEPTOR

There are at least two types of endothelin receptor, one of which has a high affinity for endothelin-1 and endothelin-2 but a low affinity for endothelin-3, while the second type of receptor has an equal affinity for all of the three isotypes of endothelin[13,14]. The former type was designated ET_A, and probably exists on the plasma membrane

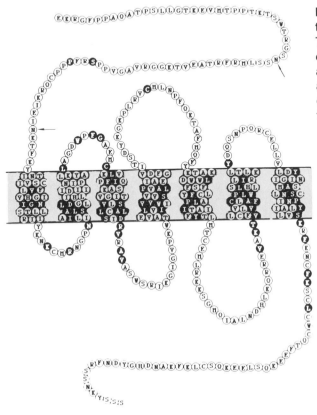

Figure 3. Schematic presentation of the endothelin receptor ET_B

Two arrows indicate two sites of extra-cellular N-linked glycosylation (Asn-33 and -70). Solid circles indicate the amino acid residues conserved in the sequ-ences of the members of this receptor family (from[1]).

of smooth muscle cells. The second type was named ET_B, and probably exists on vascular endothelial cells. Both receptors are expressed in lung, heart, brain and other tissues. In addition, ET_B is expressed in kidney and adrenal gland. Since the adrenal gland produces exclusively endothelin-3, expression of the ET_B receptor in adrenal gland is extremely interesting in terms of the function of endothelin-3 in this tissue.

The primary structures of these two receptors were determined from the sequences of the cDNA clones[13,14]. Both receptors are single-chain peptides containing seven stretches of 20 to 27 hydrophobic amino acid residues, typical of G-protein-coupled receptors of the rhodopsin superfamily. The similarity of the amino acid sequences of bovine ET_A and rat ET_B is 65%. Both receptors have two N-linked glycosylation sites in the N-terminal domain, and many serine/threonine residues in the intracellular third loop and C-terminus. The molecular masses of bovine ET_A and rat ET_B calculated from their amino acid sequences are 48516 and 49317 respectively. A schematic drawing of a structural model of ET_B is shown in Figure 3. The existence of a third type of receptor, which has a high affinity for endothelin-3 but a low affinity for endothelin-1 and endothelin-2, was predicted from studies of the binding of endothelins to various tissue membrane fractions. However, the cDNA for the third type of receptor has not yet been cloned. Since only two types of gene were detected by the cDNA probe for ET_B, this third type, if it exists, must be a quite different type of receptor.

ENDOTHELIN-STIMULATED INTRACELLULAR SIGNAL TRANSDUCTION SYSTEMS

Endothelin elicits a prompt transient increase in intracellular free calcium ions followed by a long-lasting sustained increase, not only in smooth muscle cells but also in other target cells including endothelial cells, fibroblasts, mesangial cells and cerebellar granule cells[7]. Both types of receptor, ET_A and ET_B, are involved in the endothelin-induced increase in cytosolic free calcium ions[13,14].

On the other hand, endothelin activates phospholipase C to produce inositol 1,4,5-trisphosphate and diacylglycerol[15,16]. This process is mediated by a G-protein that is insensitive to pertussis toxin[6,7]. Since inositol trisphosphate stimulates the intracellular free calcium store to release calcium ions, the endothelin-induced initial transient increase in the concentration of intracellular free calcium ion can be ascribed to the increase in production of inositol trisphosphate by endothelin. Removal of external calcium ion did not affect the initial transient increase in cytosolic calcium ion. The later sustained increase in cytosolic free calcium ion may be ascribed to influx of external calcium ion, as this later increase was not observed in a calcium-free medium (Figure 4). Moreover, nickel ion, a calcium channel blocker, abolished the endothelin-induced sustained increase in free calcium ions. Endothelin activates several types of calcium channel, including the L-type channel. This process may also be mediated by a G-protein, which is sensitive to pertussis toxin. The elevation of intracellular calcium in response to endothelin elicits long-lasting constriction in smooth muscle cells and activates nitric oxide synthesis in endothelial cells.

The mechanism for the activation of the calcium channel is unclear. One possibility is that inositol 1,3,4,5-tetrakisphosphate is involved in opening the calcium channel, as local generation of this substance has been reported to open plasma membrane calcium channels. Since the calcium channel is activated through the stimulation of

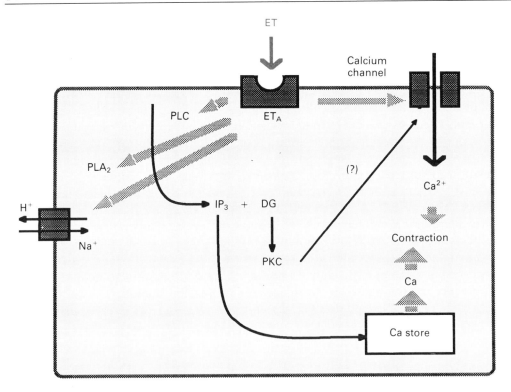

Figure 4. Schematic presentation of intracellular signal transduction pathways stimulated by endothelin-1 in smooth muscle
Abbreviations: PLC, phospholipase C; PKC, protein kinase C; PLA_2, phospholipase A_2; IP_3, inositol 1,4,5-trisphosphate; DG, diacylglycerol.

protein kinase C, this pathway may also be activated through an increase in the concentration of diacylglycerol. The endothelin-induced increase in diacylglycerol is biphasic, consisting of an early peak and a later sustained phase. The initial phase is ascribed to the diacylglycerol generated via activation of phospholipase C. However, the cause of the later phase is unclear. It may be a result of activation of phospholipase D by endothelin.

Endothelin stimulates the Na^+/H^+ antiporter, resulting in cytosolic alkalinization. The resulting intracellular alkalinization may enhance the sustained generation of diacylglycerol and may lead to a slowly developing contraction through activation of protein kinase C.

Additionally, endothelin also stimulates phospholipase A_2 to produce prostacyclin, prostaglandin E_2 and thromboxane A_2[17]. The endothelin-induced production of these prostanoids is dependent on the blood vessel employed. These prostanoids modify endothelin-induced vasoconstriction.

PHARMACOLOGICAL AND PHYSIOLOGICAL FUNCTIONS OF ENDOTHELIN

Effects of endothelin on blood flow

In isolated blood vessels, endothelin elicits slow-developing and long-lasting constriction. Intravenous bolus injection of endothelin causes an initial transient depressor response followed by a prolonged systemic pressor response in anaesthetized and conscious rats (Figure 5).

This pressor response can be entirely explained by the direct vasoconstrictor actions of endothelin on peripheral resistance vessels. Since the concentration of plasma endothelin is low, circulating endothelin is unlikely to maintain vascular tonus *in situ*. Endothelin released from endothelial cells probably acts on the underlying smooth muscle as a local hormone. A low concentration of endothelin produces a significant increase in calcium influx into cells; this calcium influx may induce *in situ* a sustained contraction of vascular smooth muscle[18]. When a bolus low dose of endothelin-1 was perfused into isolated arteries, vasodilation rather than vasoconstriction occurs at the initial step. However, endothelin ultimately elicits constriction in a dose-dependent manner. The endothelin-induced vasodilation and vasoconstriction are dependent upon the vessel employed; vasodilation is predominant in hindquarter arteries, whereas vasoconstriction is predominant in mesenteric arteries. Thus the vasoconstriction and vasodilation induced by endothelin may be mediated by different mechanisms. In mesenteric artery, endothelin-induced vasodilation can be ascribed to EDRF released from endothelium by endothelin itself. However, in hindquarter artery, endothelin-induced vasodilation seems to be mediated by a different mechanism. Endothelin-induced vasoconstriction may be mediated by activation of the ET_A receptor in smooth muscle membrane, while endothelin-induced vasodilation may be mediated by EDRF which is released through activation of the ET_B receptor on endothelial cells. Further evidence comes from the endothelin-induced response in fish: an initial transient decrease in blood pressure following bolus injection of endothelin-1 into blood vessels was observed in trout. However, trout are known to lack an EDRF-linked vasodilatory mechanism, suggesting that the initial transient depressor response induced by endothelin-1 must be ascribed to some other mechanism than EDRF released by endothelin-1.

Endothelin also causes a transient increase followed by a dramatic dose-dependent decrease in cardiac output, through an indirect mechanism involving neuronal and hormonal activity. When endothelin is infused into an animal, plasma levels of renin,

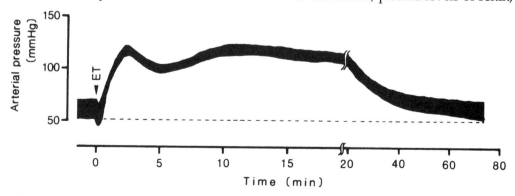

Figure 5. Effect of endothelin-1 on blood pressure
A bolus of endothelin-1 (1 ng/kg) was injected intravenously into a rat that had previously been anaesthetized with urethane and pretreated with atropine, propranolol and bunazosin. Reproduced with permission from[3].

aldosterone, atrial natriuretic peptide (ANP) and vasopressin are elevated. Endothelin directly stimulates cardiac cells to release ANP. It also stimulates the adrenal cortex to release aldosterone. However, endothelin directly inhibits the release of renin from juxtaglomerular cells. The increase in plasma renin level is probably mediated by vasoconstriction of the renal artery, which is sensitive to endothelin. However, these endothelin-released substances are unlikely to regulate vascular tonus directly.

As mentioned above, endothelin may act as a neuropeptide modulator of the cardio-vascular centre in the central nervous system. However, endothelin does not appear to affect baroreflex control of sympathetic nerve activity or heart rate.

Table 1. Pharmacological actions of endothelin

Tissue or organ	Effect
Blood vessel	Vasoconstriction and vasodilation
	Prostanoids (PGI_2, PGE_2, TXA_2) and/or EDRF release (from perfused vascular bed)
	Mitosis of endothelial, fibroblast and smooth muscle cells
	Release of tissue-type plasminogen activator and von Willebrand factor
Lymphatic vessel	Constriction
Heart	Positive inotropic and chronotropic effects
	ANP release
Kidney	
Renal artery	Vasoconstriction of renal artery
Juxtaglomerular apparatus	Inhibition of renin release (direct?) (decrease in plasma renin level)
Nephron	Diuresis and natriuresis
	Decrease in ouabain-sensitive Na^+/K^+-ATPase (inner medullary collecting duct cells)
Mesangial cell	Mitosis and contraction of mesangial cell
Adrenal gland	
Cortical glomerulosa cells	Release of aldosterone
Medullary chromaffin cells	Increase in catecholamine release
Airway muscle	Constriction
Uterus	Constriction
Liver	Glycogenolysis
Central nervous system	
Spinal cord	Increase in substance P release
Pituitary	Water balance (?)
	Increase in release of LH and FSH
	Inhibition of prolactin release
Eye	Rise in intraocular pressure
Neutrophil	Enhancement of superoxide generation elicited by fMet-Leu-Phe
	Enhancement of production of neutral protease stimulated by fMet-Leu-Phe
Stomach and ileum	Constriction

Endothelin as a cell-growth factor

Endothelin-1 stimulates DNA synthesis and proliferation of various cells including smooth muscle cells, fibroblasts, glomerular mesangial cells, endothelial cells and osteoblasts[7,15]. The potency of the mitotic activity depends on the cell employed. In general, the activity is low. However, the mitotic activity of endothelin-1 is enhanced in the presence of other factors. For example, it acts directly on 3T3 cells as a poor mitogen, but the mitogenic activity of endothelin-1 was enhanced by a low concentration of insulin-like growth factor I. A synergistic effect of endothelin with epidermal growth factor, but not with platelet-derived growth factor, was also demonstrated on DNA synthesis in cultured smooth muscle cells.

The endothelin-induced enhancement of DNA synthesis was abolished by calcium antagonists, suggesting the necessity of external calcium ion influx for this process. The mitotic effect of endothelin may be mediated by the activation of protein kinase C.

Other actions of endothelin

Endothelin has many functions not only in the cardiovascular system but in other systems as well[6,7] (Table 1). The wide distribution of endothelin-binding sites and the ubiquity of the endothelin peptide are compatible with the multifunctional nature of endothelin mentioned above.

Endothelin shows inotropic and chronotropic effects on isolated atrial muscle. Endothelin elicits constriction in almost all types of smooth muscle: airway muscle is sensitive to endothelin-1, and endothelin also acts on smooth muscle from the digestive and urogenital organs. Contraction mechanisms appear to be slightly different from tissue to tissue. In addition to uterine contraction induced by endothelin, amnion cells also produce a large amount of endothelin, suggesting that endothelin plays some role in pregnancy. In all cases, the physiological and pathophysiological effects of endothelin on smooth muscle other than vascular smooth muscle remain to be solved.

Endothelin affects kidneys in two ways. In perfusion experiments, renal blood flow and glomerular filtration rate decrease. Endothelin stimulates proliferation of mesangial cells. On the other hand, endothelin directly affects the epithelium of glomerular tubules, inducing sodium excretion. Although the physiological significance of endothelin in the kidneys is still unclear, several investigators believe that endothelin may be an aggravating factor in renal insufficiency.

As already discussed, endothelin may play an important role in the central nervous system, particularly in the pituitary. In addition, in the dorsal root ganglion of spinal cord, endothelin elicits depolarization, suggesting also that it may act as a neuropeptide.

CONCLUSIONS

- Endothelin is a potent, vasoconstrictive 21-amino-acid peptide. There are three distinct isotypes of endothelin, designated endothelin-1, endothelin-2 and endothelin-3. The endothelins are ubiquitous not only in vascular tissue but also in non-vascular tissues.

- Endothelin is produced through stimulation by many chemical factors including thrombin, cytokines and vasoactive substances.

- The endothelin receptor is also distributed widely. There are at least two types of endothelin receptor, designated ET_A and ET_B. Activated endothelin receptor stimulates phospholipase C, phospholipase A_2, the Na^+/H^+ antiporter and the opening of calcium channels to elicit a transient increase followed by a sustained increase in cytosolic free calcium ions.

- Endothelin induces constriction in all types of smooth muscle. In vascular tissue, endothelin induces a slow-rising, long-lasting contraction. An intravenous bolus injection of endothelin induces a transient depressor response at the initial step followed by a long-lasting pressor response.

- Endothelin has also hormonal activities: it stimulates the release of atrial natriuretic peptide and aldosterone, and inhibits the release of renin. In the central nervous system, endothelin may modulate the release of gonadotropins or growth hormone. It may also affect water balance.

- Endothelin has many other functions. However, the physiological and pathophysiological significance of endothelin in each tissue remains to be determined.

REFERENCES

1. Furchgott, R.F. & Zawadzki, J.V. (1980) The obligatory role of endothelial cells in the relaxation of arterial smooth muscle by acetylcholine. *Nature (London)* **288**, 373–376
2. Rubanyi, G.M. (1991) Endothelium-derived vasoactive factors in health and disease, in *Cardiovascular Significance of Endothelium-Derived Vasoactive Factors*, pp. xi–xix, Futura Publishing, Mount Kisco, NY
3. Yanagisawa, M., Kirihara, H., Kimura, S., Tomobe, Y., Kobayashi, M., Mitsui, Y., Yazaki, Y., Goto, K. & Masaki, T. (1988) A novel potent vasoconstrictor peptide produced by vascular endothelial cells. *Nature (London)* **332**, 411–415
4. Inoue, A., Yanagisawa, M., Kimura, S., Kasuya, Y., Miyauchi, T., Goto, K. & Masaki, T. (1989) The human endothelin family: three structurally and pharmacologically distinct isopeptides predicted by three separate genes. *Proc. Natl. Acad. Sci. U.S.A.* **86**, 2863–2867
5. Masaki, T., Kimura, S., Yanagisawa, M. & Goto, K. (1991) Molecular and cellular mechanism of endothelin regulation: implications for vascular function. *Circulation* **84**, 1457–1468
6. Takasaki, C., Tamiya, N., Bdolah, A., Wolleberg, Z. & Kochva E. (1988) Sarafotoxins S6: several isotoxins from *Atractaspis engaddensis* (burrowing asp) venom that affect the heart. *Toxicon* **26**, 543–548
7. Masaki, T., Yanagisawa, M. & Goto, K. (1991) Physiology and pharmacology of endothelins. *Med. Res. Rev.*, in the press
8. Koseki, C., Imai, M., Hirata, Y., Yanagisawa, M. & Masaki, T. (1989) Autoradiographic distribution in rat tissues of binding sites for endothelins: a neuropeptide? *Am. J. Physiol.* **256**, R858–R866
9. Giaid, A., Gibson, S.J., Herrero, M.T., Gentleman, S., Legon, S., Yanagisawa, M., Masaki, T., Ibrahim, N.B.N., Roberts, G.W., Rossi, M.L. & Polak, J.M. (1991) Topographical localization of endothelin mRNA and peptide immunoreactivity in neurones of the human brain. *Histochemistry* **95**, 303–314

10. Yoshizawa, T., Shinmi, O, Giad, A., Yanagisawa, M., Gibson, S.J., Kimura, S., Uchiyama, Y., Polak, J.M., Masaki, T. & Kanazawa, I. (1990) Endothelin: a novel peptide in the posterior pituitary system. *Science* **247**, 462–464

11. Stojilkovic S.S., Merelli, F., Iida, T., Krsmanovic L.K. & Catt, K.J. (1990) Endothelin stimulation of cytosolic calcium and gonadotropin secretion in anterior pituitary cells. *Science* **248**, 1663–1666

12. Okada, K., Miyazaki, Y., Takada, J., Matsuyama, K., Yamaki, T. & Yano, M. (1990) Conversion of big endothelin-1 by membrane-bound metalloendopeptidase in cultured bovine endothelial cells. *Biochem. Biophys. Res. Commun.* **171**, 1192–1198

13. Arai, H., Hori, S., Aramori, I., Ohkubo, H. & Nakanishi, S. (1990) Cloning and expression of a cDNA encoding an endothelin receptor. *Nature (London)* **348**, 730–732

14. Sakurai, T., Yanagisawa, M., Takuwa, Y., Miyazaki, H., Kimura, S., Goto, K. & Masaki, T. (1990) Cloning of a cDNA encoding a non-isopeptide-selective subtype of the endothelin receptor. *Nature (London)* **348**, 732–735

15. Takuwa, N., Takuwa, Y., Yanagisawa, M., Yamashita, K. & Masaki, T. (1989) A novel vasoactive peptide endothelin stimulates mitosis through inositol lipid turnover in Swiss 3T3 fibroblasts. *J. Biol. Chem.* **264**, 7856–7861

16. Kasuya, Y., Takuwa, Y., Yanagisawa, M., Kimura, S., Goto, K., & Masaki, T. (1989) Endothelin-1 induces vasoconstriction through two functionally distinct pathways in porcine coronary artery: contribution of phosphoinositide turnover. *Biochem. Biophys. Res. Commmun.* **161**, 1049–1055

17. Resink, T.J., Scott-Burden, T. & Buhler, F.R. (1989) Activation of phospholipase A_2 by endothelin in cultured vascular smooth muscle cells. *Biochem. Biophys. Res. Commun.* **158**, 279–286

18. Muldoon, L.L., Enslen, H., Rodland, K.D. & Magun, B.E. (1991) Stimulation of Ca^{2+} influx by endothelin-1 is subject to negative feedback by elevated intracellular Ca^{2+}. *Am. J. Physiol.* **260**, C1273–C1281

7

Roles for protein phosphorylation in synaptic transmission

Richard Rodnight and Susana T. Wofchuk

Departamento de Bioquimica, Instituto de Biociencias, Universidade Federal do Rio Grande do Sul, Rua Sarmento Leite 500, 90.050 Porto Alegre, RS, Brazil

INTRODUCTION

The term "synaptic transmission" encompasses the series of molecular events that form the basis of intercellular communication in the nervous system. The complex processes that control and modulate these events, both in the short and long term, have profound implications for all facets of brain function, from primitive autonomic functions to conscious behaviour, learning and memory. This essay will discuss the evidence that the reversible phosphorylation of synaptic proteins serves to modulate several aspects of synaptic transmission.

A protein kinase, its substrate and a protein phosphatase constitute a protein phosphorylating system. Synaptic structures are richly endowed with such systems and substantial evidence indicates that many aspects of synaptic function are regulated by the reversible phosphorylation of synaptic proteins. Phosphorylation modifies the functional properties of proteins: for example phosphorylation may stimulate or inhibit the activity of enzymes or change the conformation of non-enzymic proteins such as those that constitute ion channels or the cytoskeleton. The recent growth in our knowledge of this phenomenon is remarkable: 20 years ago less than five protein kinases were known; by 1989 the total number had risen to more than 100[1].

Table 1. Major protein kinases involved in synaptic function

Kinase	Location in synapse	Subunit structure	Typical synaptic substrates
Cyclic AMP-dependent (protein kinase A)	Pre- and post-synaptic cytosol and membrane	Tetrameric: two catalytic and two regulatory subunits	Synapsin I (head region) DARPP-32 Nicotinic acetyl-choline receptor
Ca^{2+}/calmodulin-dependent kinase I	Presynaptic cytosol	Dimeric: two major subunits	Synapsin I (head region)
Ca^{2+}/calmodulin-dependent kinase II	Presynaptic cytosol and membrane Postsynaptic densities	Dodecameric: two major and one minor subunits Proportions vary in different brain areas	Synapsin I (tail region) Tubulin Tyrosine and tryptophan hydroxylases Self (auto-phosphorylation)
Protein kinase C family	Presynaptic cytosol and membranes	Monomeric: subspecies exhibit a single poly-peptide with sequence variation	B-50 (F-1 or GAP-43) MARCKS*

*Myristoylated alanine-rich C kinase substrate, an acidic protein associated with growth in many cells (see also Table 2).

STRATEGIES OF INVESTIGATION

Two complementary strategies are being employed to investigate roles for protein phosphorylation in synaptic transmission. In the mammalian nervous system the usual approach is to use radio-labelling techniques to search for major or neurone-specific phosphoproteins, then to characterize the enzymes involved and study the regional and cellular location of the system. With this information the possible function of the system can be considered. Examples of this approach are provided by the work of Greengard's laboratory on the neuronal phosphoproteins synapsin I and DARPP-32[2]. The alternative complementary approach attempts to identify protein phosphorylation events involved in the modulation of functional responses, monitored by electrophysiology or animal behaviour. Because of its enormous complexity the application of this strategy to the mammalian brain has rarely proved possible, but it has achieved remarkable success in the simpler nervous systems of some inverte-brates such as *Aplysia*[3].

SYNAPTIC PROTEIN KINASES

The main well-characterized protein kinases encountered in synaptic structures are given in Table 1 and their typical substrates in Table 2. Some of these substrates are identified in the two-dimensional autoradiograph of rat brain tissue labelled with radioactive phosphorus shown in Figure 1.

The kinase enzymes listed in Table 1 possess a broad substrate specificity and a widespread tissue distribution, both in the nervous system and elsewhere; certain of the substrates, however, are synapse-specific. Synaptic structures also exhibit many generally less-characterized kinases often with narrow substrate specificity; these include tyrosine kinase activity[4] and kinases that specifically phosphorylate receptor proteins[5].

Of the mammalian cyclic nucleotide protein kinases, cyclic AMP-dependent protein kinase occurs in all regions of the brain, whereas the distribution of the cyclic GMP-dependent enzyme is largely restricted to the Purkinje cells of the cerebellum[2]. The properties of protein kinase A are well-documented and the synaptic enzyme does not appear to possess any special characteristics. High concentrations are found in the pre- and post-synaptic cytosol and membranes.

Of the two Ca^{2+}/calmodulin-dependent kinases (I and II) in synapses, Ca^{2+}/calmodulin-dependent kinase II is a multifunctional enzyme of particular interest[6]. The enzyme activity reflects a family of multimeric isoenzymes of high molecular mass (300–700kDa) composed of two major (α, 50kDa; β, 60kDa) and one minor (β-1, 58kDa) structurally related subunits. The various isoenzymes occur in varying proportions in different brain regions; for example in the forebrain isoenzyme the ratio of α to β is 3:1, whereas in the cerebellum the ratio is reversed (four β subunits to one α subunit). In the forebrain the kinase constitutes 1–2% of the total protein and as much as 15% of the postsynaptic density protein; it also occurs in high concentration in the presynaptic cytosol and plasma membrane. An important feature of Ca^{2+}/calmodulin-dependent kinase II is its capacity to autophosphorylate all three of its subunits. The autophosphorylation reaction initially requires calcium but in its

Table 2. Some protein kinase substrates present in synaptic structures
Proteins marked '*', or their isoforms, are found also in non-neural tissues.

Substrate	Molecular mass (kDa)	Location in synapse	Dependency of kinase(s)
Synapsin Ia and Ib	86 (a) and 80 (b)	Associated with small synaptic vesicles	Cyclic AMP Ca^{2+}/calmodulin
Synapsin IIa and IIb	74 (a) and 55 (b)	Associated with synaptic vesicles	Cyclic AMP Ca^{2+}/calmodulin
B-50 (F-1 or GAP-43)	45–48	Presynaptic plasma membrane	Ca^{2+}/lipid
DARPP-32	32	Neuronal cytosol post-synaptic to dopaminergic terminals	Cyclic AMP
MARCKS*	82–87	Presynaptic cytosol and plasma membrane	Ca^{2+}/lipid
Microtubule-associated protein 2 (MAP-2)*	>250	Dendritic post-synaptic cytosol and membrane	Cyclic AMP Ca^{2+}/calmodulin Ca^{2+}/lipid

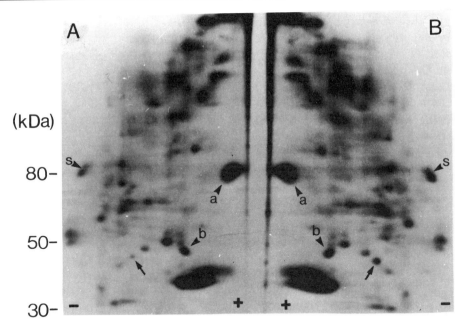

Figure 1. Autoradiograph prepared from slices of immature rat hippocampus labelled *in vitro* with radioactive phosphorus
After incubation the tissue was analysed by non-equilibrium two-dimensional electrophoresis and the gel exposed to X-ray film. Isoelectric focusing gels from two samples were mounted as mirror images on a single slab gel (8% polyacrylamide). Incubation was under control conditions in A and in the presence of glutamate in B. Several synaptic phosphoproteins can be recognized: a, MARCKS; b, B-50; s, synapsin I. The arrows point to an unknown protein whose phosphorylation was stimulated by glutamate. Adapted from Wofchuk & Rodnight[31].

phosphorylated state the enzyme loses its dependence on calcium to phosphorylate other substrates.

The other major protein kinase in synapses is protein kinase C. As is well known this enzyme is activated by calcium and diacylglycerol in the presence of phosphatidylserine. Since the latter occurs in the lipid bilayer of the plasma membrane and diacylglycerol is a product of the receptor-activated hydrolysis of phosphatidylinositol 4,5-bisphosphate by phospholipase C, protein kinase C is strategically placed to transduce one arm of the intercellular messages conveyed through receptors linked to the polyphosphatidylinositol cycle[7]. The other arm of the cycle produces inositol trisphosphate, which serves to mobilize internal calcium, part of which presumably contributes to the activation of protein kinase C.

Recent work has shown that the observed activity of protein kinase C reflects a family of kinases in which each member consists of a single common polypeptide chain with minor sequence variation[8]. Nishizuka's laboratory has identified seven subspecies, of which at least four occur in varying concentration in different regions of the mammalian brain. Perhaps the most interesting of these is the γ subspecies which is expressed solely in the brain and appears to be concentrated in presynaptic structures of the cerebral cortex, hippocampus and amygdala. The β-II subspecies is also concentrated in presynaptic terminals, but in contrast to the γ subspecies it is also expressed in non-neuronal tissues. Other subspecies appear to have non-synaptic locations. The various subspecies exhibit subtle differences in kinetic properties, suggesting that they subserve different functions.

ROLES IN SYNAPTIC TRANSMISSION

There is evidence for the modulation of synaptic transmission by protein phosphorylation in the case of the synthesis and release of neurotransmitters, receptor function, the gating of ion channels and the long-term modulation of transmission. An illustrative scheme is shown in Figure 2.

Neurotransmitter synthesis

The enzymes tyrosine hydroxylase and tryptophan hydroxylase occur in high concentration in the specific nerve terminals where they constitute the major stages limiting the rate of synthesis of the catecholamines (dopamine and noradrenaline) and the indoleamine, serotonin, respectively. The activity of these enzymes is modulated by phosphorylation. Tyrosine hydroxylase is phosphorylated by protein kinase A and by protein kinase C on the same site and by Ca^{2+}/calmodulin-dependent kinase II on a different site. Phosphorylation by protein kinases A and C stimulates the enzyme by increasing its affinity for the pterin cofactor; phosphorylation by Ca^{2+}/calmodulin-dependent kinase II, in the presence of an "activator protein", increases the $V_{max.}$ of the enzyme without affecting its cofactor affinity[9,10]. In contrast, tryptophan hydroxylase appears to be phosphorylated only by Ca^{2+}/calmodulin-dependent kinase II which also increases the $V_{max.}$ of the enzyme. These phosphorylations have been shown to occur *in situ* and may account for the increased activity of the enzymes that occurs on stimulation. However, in the case of tyrosine hydroxylase the relative importance of the cyclic AMP- or calcium-dependent kinases remains uncertain.

Neurotransmitter release

On kinetic grounds it is unlikely that protein phosphorylation is involved in the final stage of the release of neurotransmitters, the exocytotic event. This is because the delay between the entry of calcium into the terminal and release of neurotransmitter is of the order of microseconds, an interval that is probably too short to encompass a protein phosphorylation event[11]. There are, however, several lines of evidence that point to the control of earlier stages of the release process by protein kinase action.

The first concerns the neuronal phosphoprotein, synapsin I[2] (Table 2). In its dephosphorylated state synapsin I is associated through its collagen-like tail region with the surface membrane of small synaptic vesicles; phosphorylation of the tail region by Ca^{2+}/calmodulin-dependent kinase II decreases the strength of this association. In a study designed to demonstrate a role for this molecule in the release of acetylcholine, Llinas *et al.*[12] injected purified dephosphosynapsin and phosphosynapsin into the squid giant axon synapse. Dephosphosynapsin inhibited release; phosphosynapsin or heat-treated dephosphosynapsin were without effect. This result led to the hypothesis that dephosphosynapsin I crosslinks the vesicles to the cytoskeleton and/or the preterminal neuronal membrane; phosphorylation of the molecule on the

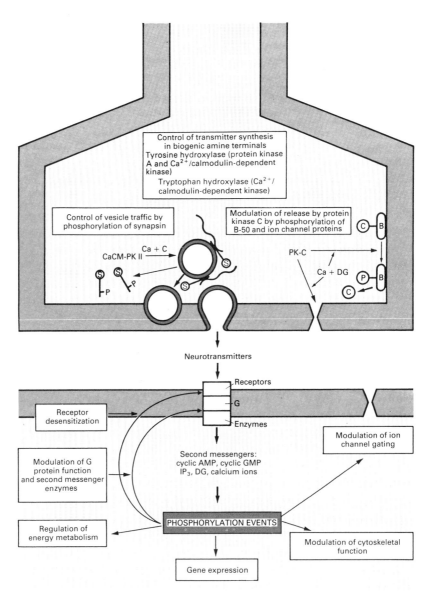

Figure 2. Schematic representation of the involvement of protein phosphorylation in synaptic transmission
In the presynaptic compartment control of vesicle movement ("vesicle traffic") is represented by a vesicle crosslinked to two actin filaments by synapsin I. Phosphorylation by Ca^{2+}/calmodulin-dependent kinase II releases synapsin and leaves the vesicle free to move to the membrane prior to exocytosis. While many details remain to be clarified there is impressive evidence for this model from Greengard's laboratory[2]. The role of protein kinase C in facilitating release may involve modulation of ion channel gating and the phosphorylation of B-50 resulting in the release of calmodulin. In the postsynaptic compartment phosphorylation events initiated by second messengers probably regulate numerous processes of which only a few are known at present. Key: S, synapsin I; B, B-50; C, calmodulin; DG, diacylglycerol; IP_3, inositol trisphosphate.

tail region by Ca^{2+}/calmodulin-dependent kinase II, consequent on calcium entry, would disrupt this link and perhaps facilitate the alignment of the vesicle in the membrane (see Figure 2). Since synapsin I is associated with all small synaptic vesicles it is reasonable to generalize this scheme to the release of neurotransmitters other than acetylcholine. In support of the model are experiments showing that synapsin I interacts with cytoskeleton elements, particularly actin filaments[13], and that depolarization of synaptosomes results in a very rapid translocation of synapsin I from the particulate fraction to the cytosol[14].

The other kinase implicated in the modulation of neurotransmitter release is protein kinase C. In a wide variety of neuronal and secretory systems, neurotransmitter or hormone release has been shown to be enhanced by exposure to protein kinase C-activating phorbol esters. These compounds, which are also tumour promoters, activate protein kinase C apparently by substituting for the natural activator, diacylglycerol. A recent study[15] demonstrated that phorbol 12,13-dibutyrate enhanced the electrically-evoked release of acetylcholine from rabbit hippocampal slices in a concentration-dependent fashion; a biologically inactive phorbol ester was without effect. The molecular basis of the modulation of neurotransmitter release by protein kinase C is unknown, but it is likely that part of the mechanism involves the control of calcium currents through the phosphorylation of ion channels by the kinase[16]. Another possibility concerns the neuronal phosphoprotein B-50[17] (Table 2; also known as GAP-43, F-1 and neuromodulin). B-50 is a membrane-associated presynaptic substrate of protein kinase C that is rapidly and transiently phosphorylated on depolarization of the terminal membrane under conditions that result in release of neurotransmitter. Antibodies to B-50, introduced into permeabilized noradrenergic synaptosomes, inhibited the release of noradrenaline in a concentration-dependent manner; control antibodies and heat-inactivated anti-B-50 antibodies were ineffective[17]. This effect of B-50 phosphorylation on neurotransmitter release may be related to the fact that dephospho-B-50, in contrast to phospho-B-50, binds calmodulin in the presence of low concentrations of calcium ions. Thus phosphorylation of the molecule following depolarization would release calmodulin and may increase its availability for the Ca^{2+}/calmodulin-dependent kinase II-catalysed phosphorylation of synapsin I (Figure 2).

Receptor functions

Many cell surface receptors are known to be phosphorylated[5]. These include receptors coupled to second messenger systems through G proteins (the α_1- and β-adrenergic receptors and the muscarinic acetylcholine receptor), a receptor which constitutes an ion channel (the nicotinic acetycholine receptor) and receptors for a variety of growth factors which are themselves protein-tyrosine kinases (e.g. the epidermal growth factor and insulin receptors). Receptor phosphorylation may be a general mechanism for the desensitization or inactivation of receptor function as a result of prolonged occupancy by agonists[18].

Of receptors clearly involved in synaptic transmission the nicotinic acetylcholine receptor from *Torpedo californica* provides an excellent model[19]. All four subunits of the receptor are phosphorylated: the γ and δ subunits by protein kinase A, the α and δ subunits by protein kinase C and the β, γ and δ subunits by an endogenous tyrosine kinase. Phosphorylation of purified preparations of the receptor subunits by protein kinase A or by endogenous tyrosine kinase activity followed by reconstitution into lipid vesicles showed that the phosphorylated form of the receptor desensitized to acetylcholine some 10 times faster than the dephosphorylated form. The interpretation of these multiple phosphorylations in terms of receptor–receptor interactions and neurotransmitter–second messenger systems has been discussed by Greengard[20].

Ion channel gating

Experiments using a variety of synaptic systems indicate that protein phosphorylation events can modulate ion channel gating[21]. Present data indicate that modulation of ionic conductance by phosphorylation is mediated by protein kinase A and by protein kinase C; the result of phosphorylation may be an increase or a decrease in conductance. The most convincing experiments have been those in which channel activity was measured in membrane patch preparations or in reconstituted lipid vesicles. For example, exposure of cell-free membrane patches from *Aplysia* sensory neurones to the catalytic subunit of protein kinase A resulted in the closure of serotonin-sensitive potassium channels[22]; in contrast, protein kinase A increased the conductance of calcium-activated potassium channels in membrane patches from the snail *Helix*[23]. With respect to protein kinase C, many examples of the modulatory actions of this kinase on ion channel gating are given in a recent review[16].

In discussing this area it is important to emphasize that only in a few cases, for example the nicotinic acetylcholine receptor channel and the voltage-dependent sodium channel[24], has the phosphorylation of proteins that constitute the ion channel been demonstrated; in most experiments the results could be equally interpreted in terms of a phosphorylation of regulatory proteins associated with the channel.

Long-term modulation of synaptic function

It is well known that brief synaptic inputs can modify the excitability of the postsynaptic neurone for prolonged periods of time, with consequential effects on all aspects of neuronal function. Several lines of evidence now support the concept that the phosphorylation of synaptic proteins plays an important role in such long-term modulations. An example is provided by the phenomenon of sensory facilitation in *Aplysia* where a single sensory stimulus may lead to behavioural modification lasting up to 1 hour. A contributing factor to the molecular mechanism underlying the behavioural sensitization appears to be the closure by phosphorylation of serotonin-sensitive K⁺ channels in the sensory neurons by protein kinase A, an event which leads to a prolonged enhancement of neurotransmitter release through secondary effects on calcium currents[25].

In the immensely complex nervous system of mammals, however, it has proved more difficult to elucidate the molecular mechanisms involved in the long-term modulation of synaptic transmission by protein phosphorylation. Most research in this area has concentrated on the phenomenon of long-term potentiation of synaptic transmission (reviewed in [26]). Long-term potentiation is an example of synaptic plasticity observed at the level of electrophysiological responses: brief high-frequency stimulation of certain afferent pathways to the hippocampus results in an increase in the postsynaptic response to stimulation of the same afferents that may last for hours

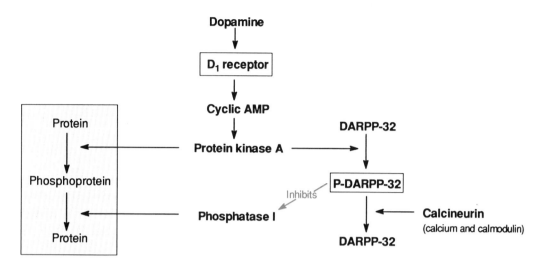

Figure 3. Scheme for the prolongation of the phosphorylated state of a protein by phosphorylation of DARPP-32
A protein phosphorylating system comprising a protein substrate, protein kinase A and protein phosphatase I is represented. Protein kinase A is shown simultaneously phosphorylating the protein and DARPP-32 as a result of a cyclic AMP message from a dopamine receptor. Phospho-DARPP-32 is then available to inhibit protein phosphatase I and prolong the physiological consequencies of the phosphorylation event. The reversal of this process may depend on the availability of Ca^{2+} and calmodulin to stimulate the dephosphorylation of phospho-DARPP-32 by activating calcineurin. Adapted from Greengard[20].

or days. The induction of long-term potentiation is associated with an enhanced release of the neurotransmitter glutamate and the postsynaptic entry of calcium promoted by N-methyl-D-aspartate receptors. While there is considerable circumstantial evidence that both protein kinase C and Ca^{2+}/calmodulin-dependent kinase II are involved in the phenomenon[27,28], the precise nature of the phosphorylation events catalysed by these enzymes remains uncertain.

Two other hypothetical mechanisms for the long-term modulation of synaptic transmission by protein phosphorylation deserve mention. The first concerns the protein kinase A substrate known as DARPP-32[2,20]. In the mammalian brain DARPP-32 occurs in high concentrations in neurones possessing dopamine receptors positively coupled to adenylate cyclase (D_1 receptors). A clue to the function of DARPP-32 was provided by the discovery that in its phosphorylated form the protein is a potent inhibitor of protein phosphatase-1, the most abundant protein phosphatase in brain; thus the cyclic AMP-mediated phosphorylation of DARPP-32 consequent on the stimulation of D_1 receptors would serve to prolong the phosphorylated state of those postsynaptic proteins that are dephosphorylated by protein phosphatase-1 (Figure 3). The time scale of the prolongation would depend on the rate of dephosphorylation of DARPP-32; this is not mediated by protein phosphatase-1, but by calcineurin, a Ca^{2+}-dependent protein phosphatase of restricted substrate specificity. This suggests complex interactions between second messenger systems and protein phosphorylation in neurones postsynaptic to dopamine pathways[20].

The second hypothetical model concerns Ca^{2+}/calmodulin-dependent kinase II. The conversion of this kinase by autophosphorylation from a calcium-dependent enzyme to a form that is independent of calcium has prompted the suggestion that the enzyme may function as a calcium-triggered "molecular switch"[6,29]. In such a scheme a transient calcium signal would result in the autophosphorylation of Ca^{2+}/calmodulin-dependent kinase II which would then continue to phosphorylate its substrates as the calcium signal faded. On this basis a molecular model for the long-term storage of information in the brain has been proposed[30].

FUTURE DIRECTIONS

Study of the molecular basis of synaptic transmission and of its modulation by protein phosphorylation is in its infancy. The complexity of the process, involving numerous interacting macromolecules, neurotransmitters and signal transduction mechanisms, presages decades of work ahead. In the mammalian brain there is virtually unlimited scope for further studies on the chemical anatomy of protein-phosphorylating systems. Modern methods of protein separation, particularly high resolution two-dimensional electrophoresis, applied to brain tissue labelled with radioactive phosphorus (Figure 1; [31]), reveal numerous unidentified neural phosphoproteins; doubtless many of these will prove to be brain-specific and part of unique protein phosphorylating systems. Once identified such phosphoproteins can be purified and antibodies raised for localization studies by immunocytochemistry; cloning, sequencing and comparison with cDNA libraries are then steps towards determining the cellular function of the protein. However, in the complex mammalian nervous system, to proceed beyond isolated systems and discover how a protein phosphorylating system modulates synaptic transmission in the intact functioning brain will require new approaches.

REFERENCES

1. Fischer, E.H. & Krebs, E.G. (1989) Commentary on "The phosphorylase *b* to *a* converting enzyme of rabbit skeletal muscle [*Biochim. Biophys. Acta* (1956) **20**, 150–157]". *Biochim. Biophys. Acta* **1000**, 297–301
2. Walaas, S.I. & Greengard, P. (1991) Protein phosphorylation and neuronal function. *Pharmacol. Rev.* **43**, 299–349
3. Kandel, E.R. & Schwartz, J.H. (1982) Molecular biology of learning. Modulation of transmitter release. *Science* **218**, 433–443
4. Hirano, A.A., Greengard, P. & Huganir, R.L. (1988) Protein tyrosine kinase activity and its endogenous substrates in brain; a subcellular and regional survey. *J. Neurochem.* **50**, 1447–1455
5. Sibley, D.R., Benovic, J.L., Caron, M.G. & Lefkowitz, R.J. (1987) Regulation of transmembrane signalling by receptor phosphorylation. *Cell* **48**, 913–922
6. Kelly, P.T. (1992) Calmodulin-dependent protein kinase II: multifunctional roles in neuronal differentiation and synaptic plasticity. *Mol. Neurobiol.*, in the press
7. Berridge M. (1986) Cell signalling through phospholipid metabolism. *J. Cell Sci. Suppl.* **4**, 137–153
8. Nishizuka, Y. (1989) The protein kinase C family: heterogeneity and its implications. *Annu. Rev. Biochem.* **58**, 31–44
9. El Mestikawy, S., Gozlan, H., Glowinski, J. & Hamon, M. (1985) Characterization of

tyrosine hydroxylase activation by K⁺-induced depolarization and/or forskolin in rat striatal slices. *J. Neurochem.* **45**, 173–184

10. Atkinson, J., Richtand, N., Schworer, C., Kuczenski, R. & Soderling T. (1987) Phosphorylation of purified rat striatal tyrosine hydroxylase by Ca^{2+}/calmodulin-dependent protein kinase II; effect of an activator protein. *J. Neurochem.* **49**, 1241–1249

11. Augustine, G.J., Charlton, M. & Smith, M.P. (1987) Calcium action in synaptic transmitter release. *Annu. Rev. Neurosci.* **10**, 633–693

12. Llinas, R., McGuinness, T.L., Leonard, C.S., Sugimori, M. & Greengard P. (1985) Intraterminal injection of synapsin I or calcium/calmodulin-dependent protein kinase II alters neurotransmitter release at the squid giant axon. *Proc. Natl. Acad. Sci. U.S.A.* **82**, 3035–3039

13. Bahler, M. & Greengard, P. (1987) Synapsin I bundles F-actin in a phosphorylation-dependent manner. *Nature (London)* **326**, 704–707

14. Sihra, T.S., Wang, J.K.T., Gorelick, F.S. & Greengard, P. (1989) Translocation of synapsin I in response to depolarisation of isolated nerve terminals. *Proc. Natl. Acad. Sci. U.S.A.* **86**, 8108–8112

15. Hertting, G. & Allgaier, C. (1988) Participation of protein kinase C and regulatory G proteins in modulation of the evoked noradrenaline release in brain. *Cell. Mol. Neurobiol.* **8**, 105–114

16. Shearman, M.S., Sekiguichi, K. & Nishizuka, Y. (1989) Modulation of ion channel activity: a key function of the protein kinase C enzyme family. *Pharmacol. Rev.* **41**, 211–237

17. Gispen, W.H., Nielander, H.B., De Graan, P.N.E., Oestreicher, A.B., Schrama, L.H. & Schotman, P. (1991) The role of the growth-associated protein B-50/GAP-43 in neuronal plasticity. *Mol. Neurobiol.* **5**, in the press

18. Huganir, R.L. & Greengard, P. (1990) Regulation of neurotransmitter receptor desensitization by protein phosphorylation. *Neuron* **5**, 555–567

19. Miles, K. & Huganir, R.L. (1988) Regulation of the nicotinic acetylcholine receptor by protein phosphorylation. *Mol. Neurobiol.* **2**, 91–124

20. Greengard, P. (1987) Neuronal phosphoproteins: mediators of signal transduction. *Mol. Neurobiol.* **1**, 81–119

21. Levitan, I.B. (1988) Modulation of ion channels in neurons and other cells. *Annu. Rev. Neurosci.* **11**, 119–136

22. Shuster, M.J., Camardo, J.S., Siegelbaum, S.A. & Kandel, E.R. (1985) Cyclic AMP-dependent protein kinase closes the serotonin sensitive K⁺-channels of *Aplysia* sensory neurons in cell-free membrane patches. *Nature (London)* **313**, 392–395

23. Ewald, D.A., Williams, A. & Levitan, I.B. (1985) Modulation of single Ca^{2+}-dependent K⁺-channel activity by protein phosphorylation. *Nature (London)* **315**, 503–505

24. Costa, M.R.C. & Catterall, W.A. (1984) Cyclic AMP-dependent phosphorylation of the α-subunit of the sodium channel in synaptic nerve-ending particles. *J. Biol. Chem.* **259**, 8210–8218

25. Castellucci, V.F., Frost, W.N., Goelet, P., Montarolo, P.G., Schacher, S., Blumenfeld, H. & Kandel, E.R. (1986) Cellular and molecular analysis of long-term sensitization in *Aplysia*. *J. Physiol. (Paris)* **81**, 349–357

26. Kennedy, M.B. (1989) Regulation of synaptic transmission in the central nervous system: long-term potentiation. *Cell* **59**, 777–787

27. Malenka, R.C., Kauer, J.A., Perkel, D.J., Mauk, M.D., Kelly, P.T., Nicoll, R.A. & Waxham, M.N. (1989) An essential role for postsynaptic calmodulin and protein kinase activity in long-term potentiation. *Nature (London)* **340**, 554–557

28. Muller, D., Buchs, P.-A., Stoppini, L. & Boddeke, H. (1992) Long-term potentiation, protein kinase C and glutamate receptors. *Mol. Neurobiol.* **5**, in the press

29. Miller, S.G. & Kennedy, M.B. (1986) Regulation of brain type II Ca^{2+}/calmodulin-dependent protein kinase by autophosphorylation: a Ca^{2+}-triggered molecular switch. *Cell* **44**, 861–870

30. Lisman, J.E. & Goldring, M.A. (1988) Feasibility of long-term storage of graded information by the Ca^{2+}/calmodulin-dependent protein kinase molecules of the postsynaptic density. *Proc. Natl. Acad. Sci. U.S.A.* **85**, 5320–5324

31. Wofchuk, S.T. & Rodnight, R. (1990) Stimulation by glutamate of the phosphorylation of two substrates of protein kinase C, B-50/GAP and MARCKS, and of ppH-47, a protein highly labelled in incubated slices from the hippocampus. *Neurosci. Res. Commun.* **6**, 135–140

RECOMMENDED FURTHER READING

Gispen, W.H. & Routtenberg, A. (eds.) (1991) Protein kinase C and its brain substrates: role in neuronal growth and plasticity. *Progress in Brain Research* volume **89**, Elsevier, Amsterdam

8

Cell death in Parkinson's disease

Susan A. Greenfield

University Department of Pharmacology, Mansfield Road, Oxford OX1 3QT, U.K.

THE CAUSES OF PARKINSON'S DISEASE

We are living longer but not necessarily better. Although most of us look forward to a fulfilling old age free of sepsis, tuberculosis and diverse illnesses of the peripheral organs, we are now faced with the prospect of degenerative disorders of the brain. Parkinson's disease[1] is one of the most common of such disorders: it is primarily characterized as a problem with movement, presenting a triad of familiar symptoms: tremor, muscle rigidity and, perhaps the most distressing of all, an inability to move easily (akinesia). Despite great advances in drug therapy which have dramatically improved the life expectancy and the life style of Parkinsonian patients, there is still no known cure. This essay is not concerned with the aetiology of Parkinson's disease, its pathological morphology, nor with the pharmacology underlying current therapy. Rather, we will be examining some of the means by which the sufferers' brain cells are killed. Unlike many other brain disturbances such as depression and schizophrenia, the site of disruption in Parkinson's disease is highly localized and well-established: there is primary loss of one specific population of cells lying deep in the mid-brain, in the "substantia nigra" (Figure 1). The neurons selectively affected are densely grouped on the upper region ("pars compacta") of the substantia nigra, they send efferent axons to the front of the brain (striatum), and contain the transmitter dopamine. We are about to examine how these dopaminergic nigrostriatal cells in particular are so prone to damage. The short answer is that no one has yet convincingly identified a single cause of Parkinson's disease, and it is looking more and more as though they never will. As our knowledge of Parkinsonian cell death increases, it appears that there are many factors involved, and that these factors can be interactive. Instead

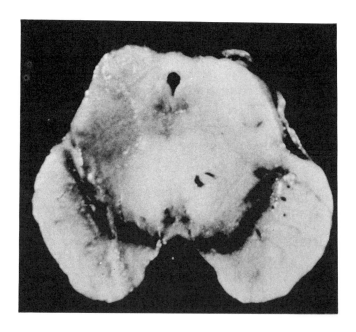

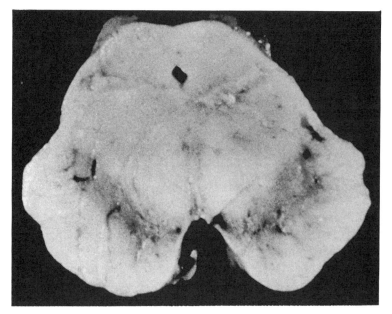

Figure 1. Sections of human midbrain showing the substantia nigra of non-Parkinsonian (above) in comparison with Parkinsonian (below) patients
In the non-Parkinsonian brain, the substantia nigra is clearly visible as a thin band of heavily pigmented cells on either side of the lower part of the brain: this region is known as the "pars compacta". The region of substantia nigra below the pars compacta is the "pars reticulata", which contains far more neuropil and more scattered cell bodies. In the Parkinsonian brain, there is a substantial loss of pigment due to degeneration of the pars compacta neurons.

of concentrating on one single cause, a more fruitful first step may thus be to somehow classify all the known relevant factors. As can be seen in Figure 2, the causes of Parkinson's disease can be readily grouped according to proximity of cause and scope of operation. The most obvious causes of cell death are "direct": sustained entry of calcium ions or excessive leakage of potassium ions can, as we shall see, have deleterious consequences for the neuron. In addition, unfettered generation of oxygen free radical species in the presence of transition metals can be very dangerous as peroxidation of the lipid plasma membrane ensues[2]. This production of radicals may occur as a result of the reduction of oxygen by its acceptance of electrons leaked from the mitochondrial respiratory chain. In aqueous solution the resultant superoxide will be converted to hydrogen peroxide by superoxide dismutase. Hydrogen peroxide will readily cross the membranes of neurons where it can react with certain transition metal ions, such as iron, to form the hydroxyl radical. This radical then attacks the fatty acid chains in lipid membranes, converting them into lipid peroxides[2]. Indeed, in Parkinsonian post mortem brains it has been shown that there is a significant increase in levels of malondialdehyde (an intermediate product of lipid peroxidation)

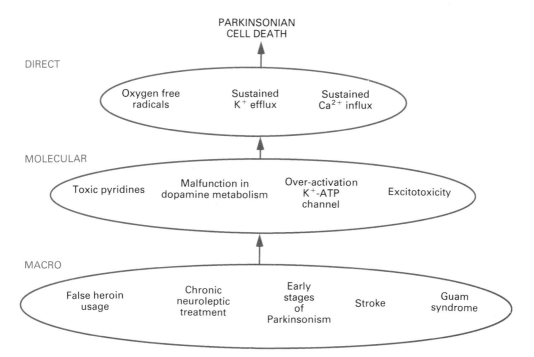

Figure 2. A classification of some of the factors leading to Parkinsonian cell death
The most direct causes would not be specific to dopaminergic cells and relate to the immediate integrity of the neuron. Lipid peroxidation by free radicals, leakiness due to excessive efflux of potassium ions or swelling of mitochondria by abnormally high levels of intracellular calcium are all lethal. These factors can be seen to arise from events at the level involving more specific transmitter and/or receptor molecules. Whereas these molecular events would be confined to particular brain regions such as the substantia nigra, they would result from a third "macro" level entailing more global disease states or drug use.

at the expense of polyunsaturated fatty acids: hence oxygen free radicals must clearly play a central role in triggering neuronal destruction[3]. However there is no reason why these immediate submolecular events should in themselves be selective for dopaminergic nigrostriatal neurons: any neuron would respond in the same way.

We can turn then to the secondary causes that resulted in these imbalances. As we shall see, transmitters and/or their receptor molecules can behave in an aberrant fashion, and hence it might be in the changes at this molecular level that we find the reason for the selective cell death seen in Parkinson's disease. But, there is still no known transmitter nor indeed receptor that is exclusive to the substantia nigra, nor is it an isolated region that functions independent of other parts of the brain. Clearly therefore there must, again in turn, have been reasons for disrupting transmitter events in the substantia nigra specifically: we can identify then yet a third "macro" level encompassing factors in other parts of the brain and in the outside world. In this essay, we will be looking at different (but not exhaustive) ways in which these three levels can link up to result in the selective death of substantia nigra cells. Since the "direct" causes are in themselves non-specific, and since the "macro" level involves many factors that are either not completely understood, nor involve the substantia nigra, we shall take as a starting point the four "molecular" issues.

TOXIC PYRIDINES

In 1978 a Californian heroin addict tried to circumvent the United States drug laws by synthesizing a substance that acted like heroin, but was chemically distinct from it. Soon he, and all those to whom he had sold his product, were appearing in clinics with a condition that was indistinguishable from idiopathic Parkinson's disease with the exception that it seemed non-progressive. Since the majority of Parkinsonian patients are over 60 years of age, these young people proved a medical mystery, until the common cause was indeed identified as a byproduct of the false heroin: this substance turned out to be a pyridine, 1-methyl-4-phenyl-1,2,3,6-tetrahydro-pyridine (MPTP)[4].

MPTP has since become the focus of intense research. It was rapidly discovered to have a highly toxic effect, in a seemingly selective fashion, on the dopaminergic nigrostriatal neurons. However, in order for this unfortunate incident to have relevance to idiopathic Parkinson's disease, the critical issue is of course how such effects might be caused. The first step towards understanding the actions of MPTP occurred when the toxic component was found to be not MPTP itself, but its metabolite 1-methyl-4-phenylpyridinium ion (MPP^+). MPTP is oxidized by monoamine oxidase B to yield the intermediate product 1-methyl-4-phenyl-2,3-dihydropyridine (MPDP). This compound is then converted to MPP^+ by monoamine oxidase B, although conversion can also occur non-enzymatically[5].

Already we now have several useful clues. First, we can identify a potentially toxic source, in that oxidation of MPTP will lead to generation of oxygen free radicals. Secondly, we can see that MPTP will only be effective in brain areas where monoamine oxidase B is present. Thirdly, within the substantia nigra monoamine oxidase B associated with dopaminergic cells is not in the neurons themselves, but distributed in non-neuronal supportive glial cells, more specifically in the star-shaped class of glial cell known as "astrocytes"[6]. Hence the state of glial cells should feature in our considerations. However, the conversion of MPTP to MPP$^+$ is only the start of the ravaging effects of the toxin.

Once converted in the astrocyte, MPP$^+$ is expelled into the extracellular space for reasons and by mechanisms that are still elusive. Nonetheless, we know that the ion is then readily and selectively taken up into catecholaminergic neurons. Here again, then, is a further form of selectivity which covers only certain populations of neurons, including of course the pars compacta neurons of the substantia nigra. The selectivity of MPP$^+$ for dopamine systems appears to stem from a more fundamental property of the toxin: it can actually be a false transmitter in catecholaminergic systems[5]. Once taken up into the cell, the actual toxicity of MPP$^+$ depends on how readily it can displace the catecholamine. In tissues where the transmitter is sequestered in robust compartments, such as in the chromaffin cells of the adrenal medulla, poisoning only occurs if storage of the catecholamine is first disrupted by an agent such as reserpine[5]. However, in nigrostriatal neurons dopamine is stored in vesicles within the striatal axon terminals and, locally in the substantia nigra itself, in the smooth endoplasmic reticulum of the somata and dendrites[7]. It is possible that the smooth endoplasmic reticulum constitutes a particularly fragile storage compartment and thus renders nigral dopaminergic neurons particularly vulnerable. On the other hand, death of neurons after MPTP appears to be from terminal to cell bodies[4]: hence dendritic storage of dopamine might not be a critical factor.

But this lethal ion has still further actions. Once inside the dopaminergic neuron, MPP$^+$ is absorbed easily into the mitochondria where it combines non-covalently with NADH dehydrogenase at the site where ubiquinone reacts. Electron transport is thus blocked, NADH oxidation and oxidative phosphorylation cease, ATP is depleted and the neuron dies[8].

The mechanisms of MPTP toxicity then include generation of free radical species both during its conversion outside the neuron and also within the cell, blockage of the electron transport chain and actions as a false (catecholaminergic) transmitter. It might still seem puzzling however, as to why the substantia nigra is particularly affected. There are two possible, not mutually exclusive, explanations. The first consideration involves the "macro" level. As we have already noted, the substantia nigra is not an isolated structure: hence any perturbation, such as that caused by MPTP, could result in swaying the delicate balance of organized interactions within the neuronal circuitry of the brain, such that highly specific long-loop feedback pathways returning to the substantia nigra have particularly damaging effects. In brief, we can shift our attention from an exclusively chemical selectivity to include an anatomical selectivity. We shall be returning to the identity and mechanism of such damaging inputs in a later section ("Excitotoxicity"); for the moment we will concentrate on the substantia nigra *per se*.

The very name "substantia nigra" describes its most conspicuous feature, that it appears black in fresh tissue. This blackness is due to the large aggregations of the pigment neuromelanin, which is not however present in the substantia nigra of animals lower down the phylogenetic chain, such as rats. This fact provides us with the second hint as to why the substantia nigra is particularly vulnerable to MPTP. A

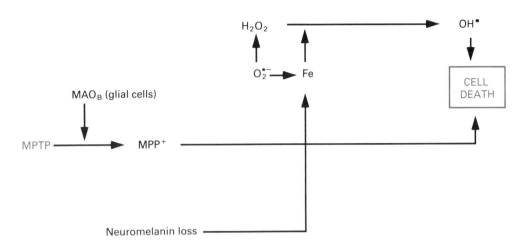

Figure 3. Relation of MPTP to cell death in the substantia nigra
MPTP is oxidized by monoamine oxidase B in glial cells to yield the ion MPP⁺. This ion can act
directly to contribute to cell death by acting as a false transmitter. A further way MPP⁺ exerts its
toxicity is via generation of oxygen free radicals resulting from oxidation of MPTP, or from
disrupting the respiratory electron transport chain by blocking NADH dehydrogenase and
oxidation of cytochromes distal to Complex 1. The presence of iron, or indeed other transition
metals, is vital to the transformation of H_2O_2 to oxygen free radicals: in turn, superoxide anion
may mobilize further iron from its sequestration with ferritin or with neuromelanin.

surprising but now well-established feature of the toxin is that it most easily destroys
the substantiae nigrae of species where melanin is present, namely primates. Melanin
then in some way could potentiate the action of MPTP so that it crosses some type
of threshold of toxicity. One possibility is that MPP⁺ is sequestered by the pigment
rather than being secreted from the neurons, as it is in other, non-nigral, tissue; the
second idea is that melanin is merely an index of catecholaminergic cells that are
particularly prone to damage. Indeed, the more melanin present in the substantia
nigra, the more vulnerable the neurons appear to be. As we shall see in the next
section, dopamine can undergo autoxidation to quinone. This quinone product then
polymerizes to form melanin: hence generation of the black pigment could be viewed
as indicative of a weakness of cellular defences to oxidative stress[9]. The fact that the
pigmented substantia nigra is particularly vulnerable to MPTP suggests that the specificity of action might lie in a homeostatic system that is already marginal.

Since oxygen free radicals appear to play such a vital intermediary step and since
we are dealing with predisposing factors rather than a single cause, we should explore
in turn the factors controlling lipid peroxidation of the membrane. The hydroxyl
radical is generated from hydrogen peroxide in the presence of ferrous ion or indeed
manganese. The presence of transition metals is therefore critical in whether or not
neuronal death occurs, and the effects of MPTP can be blocked by iron chelation.
Indeed, there is a high incidence of Parkinson's disease among manganese miners
in Chile[10]. Furthermore, in idiopathic Parkinson's disease, ferric ion is increased at
the expense of the ferrous form, suggesting an excessive amount of hydrogen peroxide.
In fact a rather vicious circle could evolve whereby iron facilitates production of free

radical species, more specifically superoxide, that catalyse the dissociation of more iron from its normally sequestered site in an erstwhile harmless complex with ferritin. On the other hand, iron in the substantia nigra can be as valuable[11] as it is dangerous[12]. Dopamine is synthesized from the precursor tyrosine by tyrosine hydroxylase, and a cofactor of this enzyme is ferrous ion; indeed, it has been demonstrated that iron administration can increase the activity of tyrosine hydroxylase in the nigrostriatal pathway by 20-fold[11]. Then again, iron deficiency or iron overload *in vivo* do not appear to affect tyrosine hydroxylase activity; there is even the possibility that iron activates the dopamine receptor[14]. However, the long latency of iron-induced super-sensitivity (weeks) exceeds the time frame[14] for the therapeutic effects of iron in Parkinson's disease[11].

We have seen so far then that many interactive factors are brought into play following exposure to toxic pyridines: some, at least, of these interactions are summarized in Figure 3. However, when James Parkinson first described the disorder which bears his name, in 1817, no one was synthesizing heroin analogues. We can turn now therefore to a more basic underlying issue, that would even have been relevant to patients of that earlier era, the fact that dopamine itself can be toxic.

MALFUNCTION OF DOPAMINE METABOLISM

Dopamine can be broken down in two ways. First, there is pre-synaptic oxidative deamination where the primary aldehyde product is metabolized further to 3,4-di-hydroxyphenylacetic acid:

dopamine + H_2O + O_2 → 3,4-dihydroxyphenylacetaldehyde + H_2O_2 + NH_3

Secondly, post-synaptic deamination can occur of the metabolite 3-*O*-methyl-dopamine to homovanillic acid:

3-*O*-methyldopamine + H_2O + O_2 → homovanillaldehyde + H_2O_2 + NH_3

or in general:

RCH_2NH_2 + H_2O + O_2 → RCHO + H_2O_2 + NH_3

Although both these reactions are well known, only relatively recently has attention swung to the fact that in both cases toxic oxygen species can be generated[13]. We come again then to generation of free radicals as a final common factor in Parkinsonian cell death, albeit by a different route to that discussed in the previous section. Clearly then, any situation that favoured an increase in dopamine metabolism might be viewed as precipitating Parkinson's disease.

One of the most common ways in which dopamine turnover can be increased is by chronic administration of a drug that blocks the dopamine receptor, such as the phenothiazines or butyrophenones. Such drugs are commonly given to schizophrenics because of their tranquilizing "anti-psychotic" action. In neurons such as those in the substantia nigra, the antipsychotic drugs block the feedback onto "autoreceptors" normally provided by the released transmitter, so that still more dopamine is produced and released. This excessive amount of dopamine will then provide further substrate for the reactions described above, leading to generation of hydrogen peroxide and subsequent production of hydroxyl radicals. We know that this chain of events is more than theoretical since the cerebrospinal fluid of patients undergoing long-term phenothiazine/butyrophenone therapy and displaying movement disorders contains abnormally high levels of free radicals, as reflected in the levels of conjugated dienes.

Another scenario in which the exuberant turnover of dopamine could have aversive consequences is the early stages of Parkinsonian degeneration itself. It is a fascinating feature of the disease that the movement disorders only appear once at least 70% of nigrostriatal neurons have been lost. Clearly then, some form of neuronal compensation must be taking place; however, if this compensation involves the extant neurons releasing more dopamine, then clearly the strategy will mis-fire. The more the remaining nigrostriatal neurons attempt to redress the balance by releasing more dopamine, the more they will be precipitating the production of potentially toxic oxygen free radicals. This unfortunate feed-forward cycle of events could account for the otherwise puzzling process of degeneration, where a sustained hostile factor would need to be continuously present.

The danger of excessive amounts of dopamine is compounded further by an ability of the transmitter to yield free radicals independent of its breakdown by synaptic monoamine oxidase B. As mentioned in the previous section, dopamine can be readily autoxidized to yield quinones and semiquinones, with simultaneous generation of superoxide anion and hydrogen peroxide. Possible modification of proteins through reaction with sulphydryl groups could add to the toxicity of these quinones. The process of autoxidation normally kept in check in the brain by two principal enzymes,

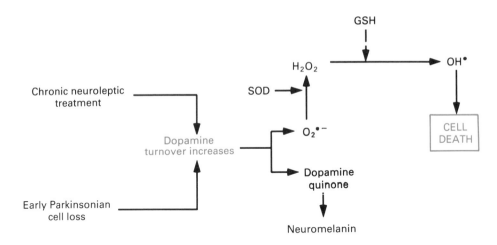

Figure 4. Examples of ways in which dopamine turnover can increase and, in turn, have pathological consequences

Blockade of dopamine receptors both presynaptically at the striatal nerve terminal and at "autoreceptors" on the perikaryon in the substantia nigra leads to a compensatory increase in dopamine turnover and release. Similarly, compensation is known to occur in the early stages of Parkinson's disease until up to 70% of cell loss: this compensation would entail an increase in dopamine metabolism in the intact cells. The excess of dopamine will lead to a high degree of autoxidation yielding oxygen free radicals that could nonetheless be offset by superoxide dismutase (SOD) and glutathione (GSH). By contrast, dopamine turnover might be normal but the autoxidation products rendered toxic by low levels of superoxide dismutase and/or GSH. Oxygen free radicals can also be formed indirectly from the redox cycling of another autoxidation product, quinone, which will polymerize to form neuromelanin.

superoxide dismutase and glutathione (GSH) peroxidase[2,13]. In other systems the enzyme catalase is also used as a cellular defence against oxidative stress, as it simply decomposes hydrogen peroxide to water and oxygen; however, in brain tissue catalase does not appear to be very active. Rather, superoxide dismutase will dismute superoxide to hydrogen peroxide, which glutathione peroxidase then uses to oxidize GSH. The potential toxicity of autoxidized dopamine is thus offset by these two enzymes. We can therefore discern still further factors at play in our consideration of malfunctions in dopamine metabolism: if superoxide dismutase, GSH or glutathione peroxidase levels are low, for example in old age, then the potentially deleterious action of dopamine itself will be more easily realised. Indeed, levels of GSH in the nigrostriatal system are lower in any event than elsewhere. An attempt at relating cell death to some of the factors affecting dopamine metabolism is shown in Figure 4. Nonetheless there is yet a further way in which an excessive dopamine could be generated, and that is by a third "macro" cause, ischaemia.

Cells can become anoxic in four different ways: hypoxic hypoxia, where there is a deficiency in the primary oxygen source; anaemia, where there is insufficient haemoglobin; histotoxic hypoxia, where the cells are unable to utilize oxygen; and finally perhaps the most common cause of anoxia in brain cells, ischaemic or stagnant hypoxia which occurs when the local blood circulation has failed, as in stroke. Transient anoxia can be highly pernicious. First, there is a collapse of normal energy-consuming cellular mechanisms, and hence released dopamine will no longer be taken up back into the neuron. Rather, this released dopamine will accumulate in the extracellular space and provide abundant substrate for autoxidation processes and increases in free radicals, as described in the previous section[13]. Secondly, as ATP levels fall, AMP will accumulate and be degraded to hypoxanthine: this hypoxanthine is normally a substrate for xanthine dehydrogenase which catalyses a reaction to uric acid. However, during anoxia/reperfusion, xanthine dehydrogenase undergoes conversion to xanthine oxidase by limited proteolysis, such that its reaction with hypoxanthine now leads to the generation of superoxide[13]. On the other hand, it is still a controversial issue whether or not this conversion of xanthine dehydrogenase is important in humans. Levels of superoxide and hydrogen peroxide are also increased in a third, less specific way during ischaemia, by enhanced mitochondrial leak. However there is a further fourth result of ischaemia that does not immediately implicate free radicals, but draws attention to the potentially lethal effect of excessive leakage of potassium ions.

THE K$^+$-ATP CHANNEL

We normally think of the flux of ions into and out of a neuron as being the result of the opening of ionic channels that are in turn ligand- or voltage-gated. Fairly recently, however, a new type of potassium channel was discovered that is primarily sensitive to metabolic state as reflected in levels of ATP. When the neuron is metabolically compromised and ATP levels are low, this "K$^+$-ATP" channel opens, allowing the efflux of potassium ions[15]. The presence of K$^+$-ATP channels in a neuron would thus indicate that neuron, if lacking energy, would have a ready means to hyperpolarize, hence become quiescent, and so be more economical in its subsequent energy

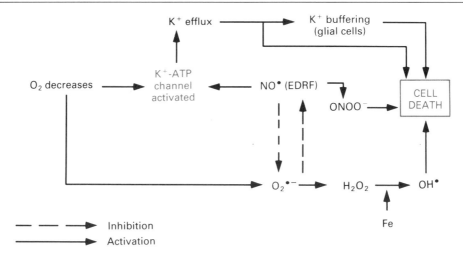

Figure 5. A pivotal role for the ATP-dependent potassium (K⁺-ATP) channel in cell death in the substantia nigra
One way the channel will be activated is during anoxia, and another by nitric oxide (NO). If opened to a marked extent, the channel will emit a large efflux of potassium ions, causing a decrease in electroresponsiveness and eventual death. This possibility would be increased if the glial cells were in any way deficient and less able to act as a spatial buffer to the extracellular potassium. Anoxia can also have a potentially lethal effect independent of the K⁺-ATP channel by enhancing generation of free radicals through enhanced accumulation of dopamine (and hence increased autoxidation) and by causing partial proteolysis of xanthine dehydrogenase to xanthine oxidase. These superoxide radicals would be in part neutralized by NO˙ free radicals, but could nonetheless be hazardous as peroxynitrite anions, which would decompose to hydroxide free radicals and nitrogen dioxide.

demands. On the other hand, that same cell would run the risk that an excessive non-physiological opening of the K⁺-ATP channel, as in clinical ischaemia, might result in such an efflux of potassium ions that the membrane became pathologically leaky: the electroresponsiveness of the neuron would then be severely curtailed. The K⁺-ATP channel thus signifies a neuron sensitive to, but at risk from, respectively physiological and pathological perturbations in energy consumption. The neurons with the highest density of K⁺-ATP channels throughout the brain are in the substantia nigra[16].

It would follow then that any substance that blocked the K⁺-ATP channel might be valuable in pathological situations where the neurons were metabolically compromised. One of the most potent and selective type of K⁺-ATP blocker is the sulphonylurea group of drugs (glibenclamide, tolbutamide), which are in established use as antidiabetic agents[15,16]. It was first reported 30 years ago that tolbutamide was efficacious in Parkinson's disease, although at that time the underlying reason was unknown. This observation nonetheless clearly illustrates that the K⁺-ATP channel is yet another factor in Parkinsonian cell death. However we can pursue this theme still further.

A secondary factor contributing to the potential power of excessive levels of extracellular potassium is determined by the non-neuronal cells mentioned briefly in our discussion of MPTP, the glia. Of the many vital functions of glia, one is to remove excessive quantities of potassium from the extracellular space — "potassium spatial buffering"[17]. Hence, in a substantia nigra where there was a malfunction of glia, ischaemia or activation of the K[+]-ATP channel by some other means might have consequences that were more dire than otherwise. More generally we can see that in any event non-neuronal cells may be intimately involved with neuronal degeneration and should thus be included in any scheme linking possible Parkinsonian pathology to a large efflux of potassium ions (Figure 5).

Since the K[+]-ATP channel is in such a pivotal position relating metabolic to electrical events, it is clearly important to explore other ways in which it might be activated. Only recently has it been discovered that nitric oxide can directly activate the K[+]-ATP channel[18], and indeed the neuroactivity of such a simple molecule is only just being realized.

NITRIC OXIDE AND EXCITOTOXICITY

Nitric oxide is now known actually to be endothelium-derived relaxant factor (EDRF), which has potent vasodilatatory actions on peripheral blood vessels. However, we also know that nitric oxide can play an important role in neuronal transmission by increasing levels of cyclic GMP through an activation of guanylate cyclase[19]. For our purposes here nitric oxide is important in a dual capacity, firstly because it can activate the K[+]-ATP channels[18] (perhaps via raised levels of cyclic GMP) and, secondly because it is itself a free radical and will thus neutralize other free radicals such as superoxide or hydroxyl. Indeed, the nitric oxide-donating substance nitroprusside has proved protective against neuronal damage by oxygen radicals[19]. On the other hand, the peroxynitrite anions formed by a reaction of nitric oxide with superoxide could decompose to yield hydroxyl radicals[19].

Endogenous nitric oxide can be synthesized in neurons during the conversion of arginine to citrulline by the cleavage of a terminal guanidino nitrogen by nitric oxide synthase. In turn, nitric oxide synthetase is activated by calmodulin which is of course dependent on the influx of calcium ions[19]. We could say then that one way to arrange for the K[+]-ATP channel to be activated, by nitric oxide, would be to implement a way for sustained entry of calcium into the nigrostriatal neuron. The most potent means by which calcium can enter a neuron is via a subtype of glutamate receptor–ion channel complex, that sensitive to N-methyl-D-aspartate, the so-called NMDA receptor[19].

There are two brain regions that both send major inputs to the substantia nigra and use glutamate as a transmitter: the pedunculopontine and subthalamic nuclei[20]. It is possible that activation of either of these pathways could lead to activation of nigral NMDA receptors[20], sustained entry of calcium[19,20], production of nitric oxide[19] and activation of the K[+]-ATP channel[18].

It is obvious however that activation of glutamate afferents to the substantia nigra would not just result in the opening of an ion channel; since we are dealing with an initial event as basic as the entry of calcium, we should consider other consequences also, many of them bad. Calcium is well known to be the trigger for most intracellular signalling including the activation of calcium-dependent lipases and proteases; however, the ion can also have a direct deleterious effect by swelling the mitochondria

Table 1. Evidence for a role of glutamate in the degeneration of dopaminergic neurons in the substantia nigra.

The substantia nigra receives two glutamate inputs, from the pedunculopontine nucleus and from the subthalamus. The glutamate released from these afferent fibres would act on post-synaptic NMDA receptors within the subtantia nigra.

1.	β-N-Methylamino-L-alanine (Guam syndrome) mimics NMDA agonists[21]
2.	Chlorokynurate (antagonist at the glycine allosteric site of the NMDA receptor) protects against NMDA-induced degeneration (in hippocampus and striatum)[23]
3.	Subthalamo-nigral pathway (glutaminergic) is abnormally active following MPTP[20]
4.	NMDA antagonist enhances locomotor activity in monoamine-depleted mice[24]
5.	Lesion of subthalamus, and hence removal of a glutamate input to the subtantia nigra, ameliorates animal models of Parkinsonism[25]
6.	Within the substantia nigra, glutamate receptor blockers (NMDA antagonists) protect against MPP⁺ neurotoxicity[26]

and preventing oxidative phosphorylation. It follows then that although calcium might first excite a neuron by depolarization, it could then be highly toxic; any substance causing the prolonged entry of calcium would thus be "excitotoxic". Sadly, there is an example of excitotoxicity at work within a particular population of people, the inhabitants of the Pacific island of Guam[21].

Guam has long been infamous for an extraordinarily high incidence of neuronal degenerative disorders, including Parkinson's disease. For many years this "Guam syndrome" proved a puzzle until a type of flour was identified that was exclusive to the island diet. This flour was ground from the cycad seed, whence the toxic agent was eventually identified as β-N-methylamino-L-alanine[21]. β-N-Methylamino-L-alanine seemed to act as an excitotoxin, and indeed its structure bears a striking similarity to that of N-methyl-D-aspartate in the presence of bicarbonate[21]. "Guam syndrome" then can be seen to provide evidence that, more widely, the glutamate inputs to the substantia nigra might precipitate degeneration if they were overactive. This idea is supported by diverse other observations (Table 1). A particularly important issue is the "physiological" factor involved in the complex neuronal circuitry, i.e. the balance of activation and inhibition. As nigral neurons become silenced as with MPTP, the subthalamo-nigral pathway becomes more excited[20]. Hence more glutamate is released as a result of the "long-loop" anatomy and physiology mentioned earlier.

Not only is the 'physiological' status of the subthalamo-nigral pathway important, namely its activity or otherwise in the control of movement, but equally vital is the very nature of the excitotoxic transmitter used, glutamate. The specificity of glutamate is relevant not only in terms of its target, the powerful NMDA receptor–calcium channel complex, but also in terms of the very metabolism of the molecule. Glutamate is of course the precursor of the inhibitory transmitter γ-aminobutyrate (GABA), and indeed there is a γ-aminobutyrate-containing projection from the striatum and globus pallidus, to the substantia nigra. The hyperpolarizing effects of γ-aminobutyrate will have an antagonistic, opposing action to that of glutamate due to a particular feature of the NMDA receptor: the NMDA receptor can only be activated after a prolonged

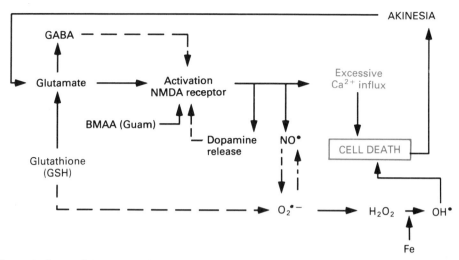

Figure 6. A possible mechanism for excitotoxicity leading to cell death in the substantia nigra

Excitotoxicity results from sustained stimulation of the NMDA receptor: the resultant large influx of calcium ions can *inter alia* swell the mitochondria and prevent oxidative phosphorylation and manufacture of ATP. The entry of calcium will also trigger manufacture of NO•, leading to a potential protection against oxygen free radicals, but also further activation of the K^+-ATP channel. A third consequence of NMDA receptor activation would be to cause local release of dopamine. Dopamine, as γ-aminobutyrate, will have a depressant, hyperpolarizing action on the neuron and thus abolish a prerequisite for NMDA receptor activation, depolarization. On the other hand, the receptor could be activated under the depolarizing actions of the γ-amino-butyrate precursor glutamate and its analogue β-N-methylamino-L-alanine (BMAA). Glutamate avalability will depend on its precursor glutathione (GSH), which itself will help scavenge oxygen free radicals. A further factor increasing the release of glutamate will be the long-loop feedback resulting from the initial cell death (see Table 1), which will have caused poverty of movement (akinesia).

period of depolarization has removed the otherwise blocking action of magnesium ions[19]. Hence, the action of γ-aminobutyrate would be to eliminate the prerequisite for the powerful potential of glutamate to be realised. In addition however, glutamate is also metabolically cycled with glutathione (GSH) by means of the enzymes GSH synthetase, γ-glutamylcysteine synthetase and γ-glutamyltransferase[22]. It is possible therefore that glutamate levels in the substantia nigra might also influence levels of GSH discussed earlier. The very chemical nature of glutamate therefore will curb its own actions. Some of the biochemical interactions involving excitotoxicity are shown in Figure 6.

CONCLUSIONS

We have seen that many factors can be involved in Parkinsonian degeneration, but that these factors can be classified in a way whereby at least four distinct threads (Figures 3, 4, 5 and 6) can be traced from the "macro" to the "direct" submolecular level. On the other hand it would be erroneous to assume that each of the four scenarios described here should be considered independently. Figure 7 may seem

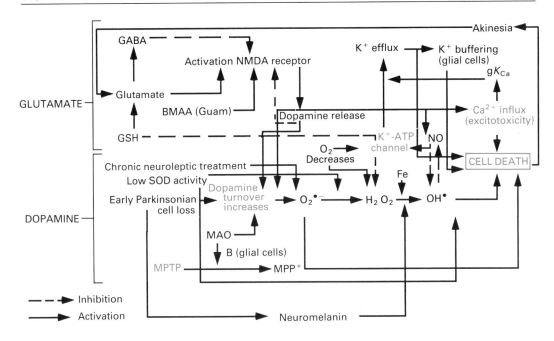

Figure 7. A summary scheme of some of the mechanisms possibly involved in cell death in the substantia nigra
These mechanisms involve primarily dopamine (toxic pyridines, Figure 3; malfunction in dopamine metabolism, Figure 4) or glutamate (overactivation of the K^+-ATP channel, Figure 5; excitotoxicity, Figure 6).

highly complex; however, it is a simple amalgam of the preceding four figures, with no additional information added. There are several important concepts to grasp regarding this summary scheme. First, the interactions shown are far from exhaustive. Not only must there still be many still unknown biochemical mechanisms at play, but even if we restrict ourselves to the immediate scope of dopaminergic nigrostriatal neurons, there are more features that could have been included. For example, these cells have the unusual property of releasing dopamine not only from striatal axon terminals, but also locally, from dendrites within the substantia nigra itself. In addition a soluble form of acetylcholinesterase is secreted that has a "modulatory" action independent of cholinergic transmission. It would be hard to imagine that these non-classical phenomena[7] did not in some way impinge on the scheme of events outlined in Figure 7, although we do not yet know how. Furthermore, the events outlined concern only two transmitter systems, glutamate and dopamine. The nigra is qualitatively and quantitatively rich in diverse neurochemicals — it has actually been referred to as "the treasure trove of the pharmacologist" — and we should therefore consider the two systems described as merely the tip of the iceberg in illustrating biochemical interactions related to nigral cell death.

Secondly, in the substantia nigra the theme that emerges is one of very fine balance between diverse transmitter systems. There is no new villainous substance in our scheme; all are necessary for the normal functioning of the substantia nigra, and rather it is their relationships that have become aberrant. The factors involved in

Parkinsonian degeneration resemble a house of cards: remove any component, and the whole edifice falls because there is a high degree of interdependence. Furthermore, these relationships are not merely biochemical but involve a consideration of the anatomical pathways and the physiological state of those pathways. It is the contingency of particular biochemical, anatomical and physiological features that give the substantia nigra its individual signature and perhaps render it so selectively prone to Parkinsonian damage. On the other hand we must remember that other areas (e.g. the forebrain) and other systems (e.g. dopamine) are involved in disorders such as Alzheimer's disease and schizophrenia respectively. Perhaps in these conditions also there are comparable multidisciplinary neuronal "signatures" not yet fully revealed. It is clearly very misleading therefore to consider central nervous system disease in terms of only one neurotransmitter system.

Our final and third consideration is perhaps the most pressing: how do the conclusions we are reaching help in treating Parkinson's disease? In the short term, the outlook might seem gloomy in that there is clearly no simple perturbation that can be reversed, no single substance to be destroyed or replaced. On the other hand, the multifactorial nature of the situation might in the longer term provide greater grounds for optimism. If there are many factors that can be exploited pharmaceutically, then three main problems with drug therapy, side effects, habituation and non-specificity, could be minimized. First, if there were side effects or habituation to any particular drug, then it could be replaced for a time by an alternative drug working in a completely different way. Secondly, we can see that there is ever-expanding scope for the design of more drug "cocktails" where each individual drug is given at a dose rendering it relatively ineffectual in its own right (and thus reducing the risks of side-effects and non-specific action), but where the combination of drugs exploits specific biochemical relationships, which occur in specific brain regions, of the type described here. Knowledge of the neurochemical mechanisms of homeostasis in the substantia nigra, and more significantly, an understanding of how these mechanisms form complex interactions, will provide us with increasing insight into the elusive link between the mental and the mechanical, from chemical events to the generation of normal and abnormal movement.

REFERENCES

1. Factor, S.A. & Weiner, W.J. (1988) The current clinical picture of Parkinson's disease, in *Progress in Parkinson Research* (Hefti, F. & Weiner, W.J., eds.), pp. 1–10, Plenum Press, New York and London
2. Halliwell, B. & Gutteridge, J.M.C. (1985) Oxygen radicals and the nervous system. *Trends Neurosci.* **8**, 22–26
3. Dexter, D., Carter, C., Agid, F., Agid, Y., Lees, A.J., Jenner, P. & Marsden, C.D. (1986) Lipid peroxidation as cause of nigral cell death in Parkinson's disease. *Lancet* **ii**, 639–640
4. McCrodden, J.M., Tipton, K.F. & Sullivan, J.P. (1990). The neurotoxicity of MPTP and the relevance to Parkinson's disease. *Pharmacol. Toxicol.* **67**, 8–13
5. Reinhard, J.F., Jr., Diliberto, E.J. & Daniels, A.J. (1988) MPP$^+$ (1-methyl-4-phenyl-pyridinium), a neurotoxin formed *in situ* from MPTP, is a false adrenergic transmitter, inactivated through sequestration in the catecholamine storage vesicle, in *Neurotoxins in Neurochemistry* (Dolly, J.O., ed.), pp. 132–147, Ellis Horwood, Chichester
6. Brooks, W.J., Jarvis, M.F. & Wagner, G.C. (1989) Astrocytes as a primary locus for the conversion of MPTP into MPP$^+$. *J. Neural Transm.* **76**, 1–12

7. Greenfield, S.A. (1985) The significance of dendritic release of transmitter and protein in the substantia nigra. *Neurochem. Int.* **7**, 887–901

8. Sanchez-Ramos, J.R., Hefti, F., Hollinden, G.E., Sick, T.J. & Rosenthal, M. (1988) Mechanisms of MPP$^+$ neurotoxicity: oxyradical and mitochondrial inhibition hypotheses, in *Progress in Parkinson Research* (Hefti, F. & Weiner, W.J., eds.), pp. 145–152, Plenum Press, New York and London

9. Graham, D.G. (1978) Oxidative pathways for catecholamines in the genesis of neuro-melanin and cytotoxic quinones. *Mol. Pharmacol.* **14**, 633–634

10. Bencko, V. Gikrt, M (1984) Manganese: a review of occupational and environmental toxicology. *J. Hyg. Epidemiol. Microbiol. Immunol.* **28**, 139–168

11. Birkmayer, W. & Birkmayer, J.G.D. (1989) Iron therapy in Parkinson's disease: stimulation of endogenous presynaptic L-DOPA biosynthesis by the iron compound oxyferriscorbone, in *Early Diagnosis and Preventative Therapy in Parkinson's Disease* (Przuntek, H. & Riederer, P., eds.), pp. 323–327, Springer-Verlag, Wien, New York

12. Sofu, E., Riechler, P., Heinsen, H., Beckmann, H., Reynolds, G.P. & Hebenstreit, G. (1988) Increased iron(III) and total iron content in post mortem substantia nigra of Parkinsonian brain. *J. Neural Transm.* **74**, 13P

13. Braughler, J.M. & Hall, E.D. (1989) Central nervous system trauma and stroke. *Free Radical Biol. Med.* **6**, 289–301

14. Youdim, M.B.H. (1989) Dopaminergic neurotransmission and status of brain iron, in *Early Diagnosis and Preventative Therapy in Parkinson's Disease* (Przuntek, H. & Riederer, P., eds.), pp. 151–160, Springer-Verlag, Wien, New York

15. Ashcroft, F.M. (1988) Adenosine 5′-triphosphate-sensitive potassium channels. *Annu. Rev. Neurosci.* **11**, 97–118

16. Treherne, J.M. & Ashford, M.L.J. (1991) The regional distribution of sulphonylurea binding sites in rat brain. *Neuroscience* **40**, 523–532

17. Kimelberg, H.K. & Norenberg, M.D. (1989) Astrocytes. *Sci. Am.* **260:4**, 66–76

18. Standen, N.B., Quayle, J.M., Davies, N.W., Brayden, J.E., Huang, Y. & Nelson, M.T. (1989) Hyperpolarizing vasodilators activate ATP-sensitive K$^+$ channels in arterial smooth muscle. *Science* **245**, 177–180

19. Garthwaite, J. (1991) Glutamate, nitric oxide and cell–cell signalling in the nervous system. *Trends Neurosci.* **14**, 60–67

20. Klockgether, T. & Turski, L. (1989) Excitatory amino acids and the basal ganglia: implications for the therapy of Parkinson's disease. *Trends Neurosci.* **12**, 285–286

21. Spencer, P.S., Nunn, P.B., Hugon, J., Lidolph, A.C., Ross, S.M., Roy, D.N. & Robertson, R.C. (1987) Guam ameyotrophic lateral sclerosis–Parkinsonism–dementia linked to plant excitant neurotoxin. *Science* **237**, 517–522

22. Proud, V.K., Hsia, Y.E. & Wolf, B. (1989) Disorders of amino acid metabolism, in *Basic Neurochemistry* (Siegel, G., Agranoff, B., Albers, R.W. & Motinoff, P., eds.), pp. 733–763, Raven Press, New York

23. Foster, A.C., Willis, C.L. & Tridgett, R. (1989) Protection against *N*-methyl-D-aspartate receptor-mediated neuronal degeneration in rat brain by 7-chlorokynurenate and 3-amino-1-hydroxypyrrolid-2-one, antagonists at the allosteric site for glycine. *Eur. J. Neurosci.* **2**, 270–277

24. Carlsson, M. & Carlsson, A. (1989) The NMDA antagonist mk-801 causes marked loco-motor stimulation in monoamine-depleted mice. *J. Neural Transm.* **75**, 221–226

25. Bergman, H., Wichmann, & Delong, M.R. (1990) Reversal of experimental Parkinsonism by lesions of the subthalamic nucleus. *Science* **249**, 1436–1438

26. Turski, L., Bressler, K., Rettig, K.-J., Loschmann, P.-A. & Wachtel, H. (1991) Protection of substantia nigra from MPP$^+$ neurotoxicity by *N*-methyl-D-aspartate antagonists. *Nature (London)* **349**, 414–418

9

Wither PQQ

William S. McIntire

Molecular Biology Division, U. S. Department of Veterans Affairs Medical Center, San Francisco, CA 94121, and the Department of Biochemistry and Biophysics and the Department of Anesthesia, University of California, San Francisco, CA 94143, U.S.A.

WHAT IS PQQ?

PQQ is an oxidation–reduction cofactor found in a variety of bacterial enzymes. It was given the trivial name methoxatin by its discoverers, because it was originally isolated from methanol dehydrogenase from the methylotrophic soil bacterium *Pseudomonas* TP1[1]. (Methylotrophic organisms are so called because they can grow on one-carbon compounds like methanol, methylamine, formate, CN^-, etc.) The systematic name for this cofactor is 4,5-dihydro-4,5-dioxo-1*H*-pyrrolo[2,3-*f*]quinoline-2,7,9-tricarboxylic acid. The cofactor consists of fused pyrrole and quinoline rings, hence the name pyrroloquinoline quinone and the acronym PQQ (Figure 1). Proteins containing this prosthetic group are called quinoproteins by analogy to flavoproteins. Unless otherwise noted, whenever PQQ is mentioned in the treatise, read 2,7,9-tricarboxy-PQQ. This distinction is important because other derivatives may exist in nature, and various derivatives have been studied in model systems.

Although only 13 years have past since the discovery of PQQ, an active field has emerged around the study of the chemical, biochemical, bioenergetic, mechanistic, biosynthetic, and molecular biological aspects of this prosthetic group. In addition to methanol dehydrogenase, PQQ is also the noncovalently-bound redox cofactor of several other classes of enzymes in bacteria (Table 1). In general, all are Gram-negative organisms, and until recently all the confirmed quinoproteins of this type were thought to oxidize primary or secondary alcohols, and/or *gem*-diols. In 1991, David J. Hopper and colleagues reported that lupinine hydroxylase contains noncovalently-bound PQQ as its redox cofactor. This enzyme is more properly an amine dehydrogenase[2] (Table 1).

All these enzymes extract two electrons from their respective substrates. The electrons are then passed into the membrane electron transport chain in the bacteria. The

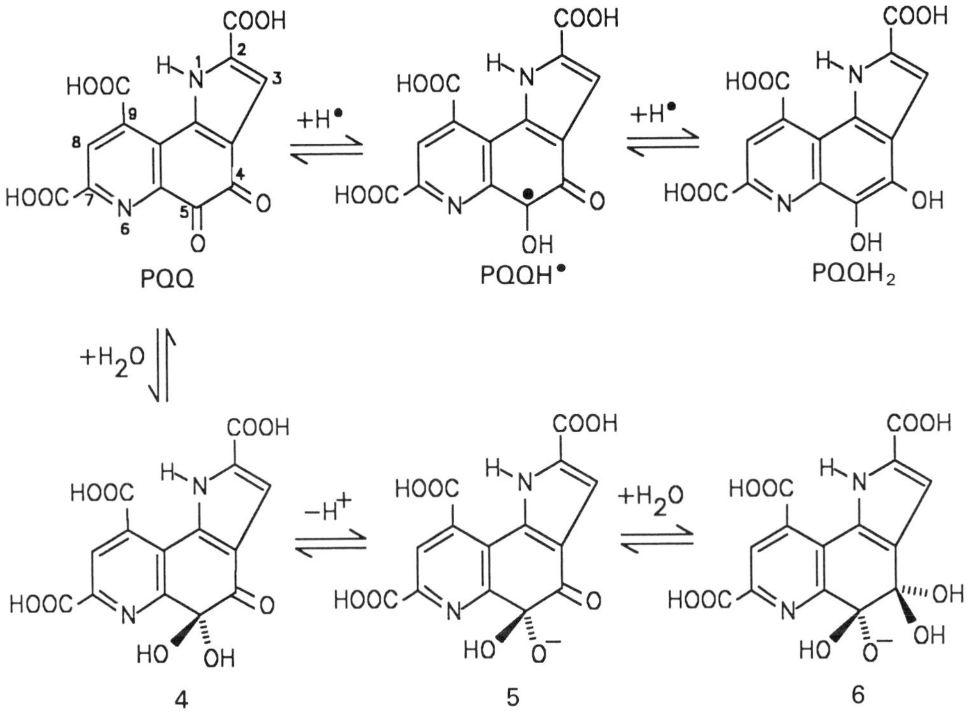

Figure 1. Redox and hydration states of 2,7,9-tricarboxy-PQQ

The upper portion of the scheme shows the one-electron reductions of PQQ (in the form of $H^\bullet$ $\equiv H^+ + e^-$). $PQQH^\bullet$ is the semiquinone radical form of PQQ. The electron is shown localized at C-5, when in fact it is delocalized throughout the tricyclic aromatic system. $PQQH_2$ is the dihydroquinone or quinol form of PQQ. Both reduced forms of PQQ can be deprotonated: $PQQH^\bullet \Leftrightarrow PQQ^{\bullet-} + H^+$; $PQQH_2 \Leftrightarrow PQQH^- + H^+ \Leftrightarrow PQQ^{2-} + H^+$. Structure 4 is the 5-hydrated form of PQQ (PQQ-HOH), and structure 5 is the pseudo-base that forms at high pH. At even higher pH the 4-position is hydrated (structure 6).

enzymes are either soluble periplasmic proteins or bound on the periplasmic face of the cytoplasmic membrane. The soluble quinoprotein couples to the electron transport chain via soluble *b*- or *c*-type cytochromes. Membrane-bound glucose dehydrogenases from *Escherichia coli*, *Zymomonas mobilis*, *Acinetobacter calcoaceticus* and *Gluconobacter suboxydans*, and membrane-bound methanol dehydrogenase from the latter organism, couple via ubiquinone.

Some of the (*mox*) genes responsible for methanol oxidation have been cloned from several methylotrophs. The sequences for the structural genes for several forms of glucose dehydrogenase and methanol dehydrogenase are known. There is a conserved consensus sequence for all of these, which may be part of the PQQ binding domain[3,4]. Other information on these enzymes can be found in several recent reviews[2,5,6].

2,7,9-Tricarboxy-PQQ has been chemically synthesized by six different methods and is commercially available. Several decarboxylated derivatives of PQQ have also been synthesized. PQQ has high reactivity towards nucleophiles, with its 5-position being the most electrophilic centre. It is attacked by water, alcohols, ammonia, alkyl and

Table 1. Quinoproteins containing noncovalently bound PQQ

Key: [a]soluble periplasmic enzymes, [b]a quinocytochrome c, [c] bound on the periplasmic face of the cytoplasmic membrane[4], [d]apparently soluble, [e]most if not all of the PQQ-containing alcohol dehydrogenases also have aldehyde dehydrogenase activity, [f]membrane-bound.

Quinoprotein and source	Reaction

Methanol dehydrogenase [a]
    Numerous Gram-negative
    methylotrophs
	$CH_3OH \rightarrow CH_2=O$ (formaldehyde)

Alcohol dehydrogenase
    Pseudomonas aeruginosa
    Pseudomonas putida
    Acetobacter polyoxogenes [b,c]
    Acetobacter aceti [b,c]
    Gluconobacter sp.[b,c]
    Comamonas testosteroni [b,d]
	$RCH_2OH \rightarrow RCH=O$ (aldehyde)

Aldehyde dehydrogenase [e]
    Gluconobacter suboxydans [b,c]
    Acetobacter aceti [b,c]
    Hyphomicrobium sp.
    Acetobacter rancens [f]
    Other Gram-positive bacteria
	$\overset{H_2O}{RCH=O \rightarrow RCH(OH)_2 \rightarrow RCOOH}$

Glucose dehydrogenase
    Acinetobacter calcoaceticus [d]
    Numerous Gram-negative bacteria [c]
	$HOCH_2(CH_2OH)_4CH=O \rightarrow HOCH_2(CH_2OH)_4COOH$

Glycerol dehydrogenase *Gluconobacter industrius* [f]	$HOCH_2CHOHCH_2OH \rightarrow HOCH_2COCH_2OH$
Lupinine hydroxylase *Pseudomonas putida* [b]	
Quinate dehydrogenase Several Gram-negative bacteria	

Polyethylene glycol dehydrogenase
    Synergistic mixture of
    Flavobacterium sp. and
    Pseudomonas sp.
	$H(OCH_2CH_2)_xOH \rightarrow H(OCH_2CH_2)_{x-1}OCH_2CH=O$

Poly(vinyl alcohol) dehydrogenase *Pseudomonas* sp. [f]	$-CH_2CHOHCH_2CHOHCH_2- \rightarrow -CH_2CHOCH_2CHOHCH_2-$

aryl amines, diamines, amino acids, acetone, aminoguanidine, urea, cyanide, sulphite, thiols, nitroalkanes, dihydronicotine analogues (hydride), borate, phenylene diamine, 2,3-diaminonaphthalene, and carbonyl reagents such as alkyl- and phenyl-hydrazines, semicarbazide and 3-methyl-2-benzothiazolinone hydrazone[1,2,5,7–11].

Figure 1 depicts the redox states of PQQ. The two-electron-reduced state ($PQQH_2$) is easily obtained by anaerobic reduction with $NaBH_4$. A four-electron-reduced form is only obtained on aerobic exposure of PQQ to excessive amounts of $NaBH_4$. The one-electron-reduced species ($PQQH^{\bullet-}$) forms by comproportionation of PQQ and $PQQH_2$ at high pH. A complete analysis of these interconversions for 2,7-dicarboxy-PQQ has been published[12]. Reported pK_a values are: 10.7 for PQQ-HOH⇔PQQ-HO$^-$ + H$^+$ (10.7 for 2,7-dicarboxy-PQQ-HOH); 8.4–8.5 for $PQQH_2$⇔$PQQH^-$ + H$^+$ (9.3 for 2,7-dicarboxy-$PQQH_2$); 8.7 for $PQQH^{\bullet-}$⇔$PQQ^{\bullet-}$ + H$^+$. The pH-dependent change of the values of the two-electron midpoint potential, E_m, of PQQ is displayed in Figure 2[2,10].

PQQ also reacts with NH_3 to produce the 5-iminoquinone ($PQQ-NH_2$). A pH titration of PQQ in the presence of NH_4^+ results in a dissociation constant for NH_3 of 41 mM, and two pK_a values of 9.1 and 11.9. E_m versus pH plots for PQQ with and without NH_4^+ are nearly identical below pH 7 (Figure 2), since at low pH very little free NH_3 is present, so very little 5-imino-PQQ forms.

Reaction with amines, hydrazines, thiols, nitroalkanes, dihydronicotinamide derivative, alcohols and aldehydes at the 5-position leads to the oxidation of these com-

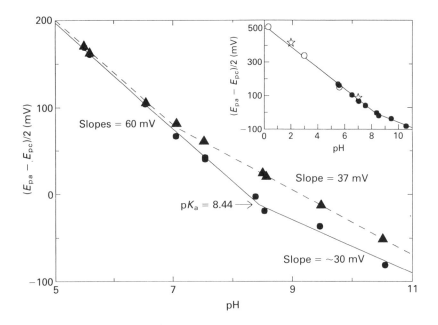

Figure 2. Plot of E_m of PQQ versus pH
E_m values were determined by cyclic voltammetry by the author. The solid line and circles are for data collected in the absence of NH_4^+, whereas the broken line and triangles are for data collected in the presence of this cation. Inset: the solid circles are as before and the open symbols show E_m data from other published studies[12].

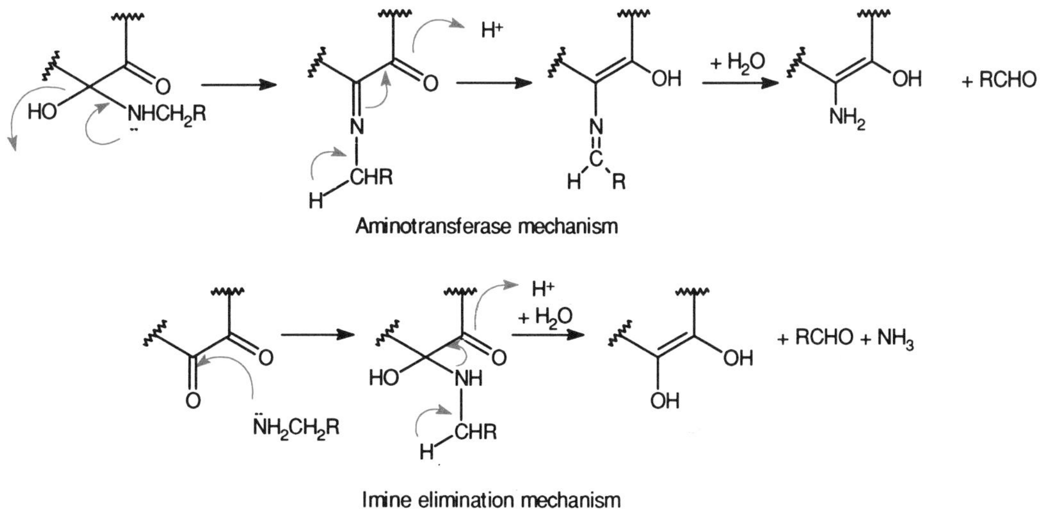

Figure 3. Proposed mechanisms of amine oxidation by *o*-quinones (e.g. PQQ) in solution

pounds and the reduction of PQQ[10]. The redox reactions of amines with PQQ have been the most actively studied because it was believe that this cofactor is the catalytic group in several amine-oxidizing enzymes. Two basic mechanisms (Figure 3) operate in the oxidation of alkyl- or aryl-amines. The aminotransferase mechanism results in the anaerobic formation of the aminophenol of PQQ. The second mechanism can be called an imine elimination mechanism, and produces PQQH$_2$. For benzylamine in aqueous solution, the aminotransferase mechanism predominates below pH 10, whereas above this pH the two mechanisms compete on an equal basis[8].

The biosynthesis of PQQ has also been actively investigated. Early speculation that PQQ is derived from tyrosine and glutamate via a multistep process has been born out[13]. Unfortunately, the nature of the enzymic processes leading from these amino acids to PQQ is unknown, although a pathway has been postulated[2,5]. Several PQQ biosynthetic genes have been identified and cloned from a variety of organisms. It would appear that, minimally, six genes (enzymes) are required for the production of PQQ[14,15].

It has been known for some time that numerous bacteria produce more PQQ than is required for endogenous quinoproteins, and excess PQQ is secreted into the medium[16]. Since methylotrophs often live in symbiosis with other organisms, the former may secret PQQ for use by symbiotic partners, in turn getting other essential nutrients from these partners[10].

The belief that several important mammalian amine oxidases contain PQQ (Table 2) prompted investigations of the nutritional role of this quinone in mice, rats and chicks[17]. Enteric bacteria do not produce PQQ, so the only source of this cofactor is dietary. Young female mice deprived of PQQ either failed to produce litters or cannibalized their young at birth. Animals deprived of PQQ suffer a number of serious abnormalities, many of them seemingly directly related to defects in connective tissue formation. (Lysyl oxidase is essential for the proper cross-linking of collagen and

Table 2. Enzymes thought, before 1990, to contain covalently bound PQQ[2,18,21]

Enzyme	Typical source	"Other prosthetic groups"
Alkylnitrile hydratase	*Brevibacterium* R312 *Rhodococcus* N-774 *Pseudomonas chlororaphis* B23	Low-spin non-haem Fe(III)
Amine dehydrogenase	*Pseuodomonas* and *Alcaligenes* species	*c*-type cytochrome
Choline dehydrognase	Mammalian liver mitochondria	Noncovalently-bound FAD
Dopamine β-hydroxylase	Bovine adrenal medulla	Type 2 copper and ascorbate
Dopa decarboxylase	Pig kidney	Pyridoxal phosphate
Galactose oxidase	*Dactylium dendroides*	Type 2 copper
Glutamic acid decarboxylase	*Escherichia coli* ATCC 11246	Pyridoxal phosphate
Laccase	Fungi	Type 1 and type 2 copper
Lipoxygenase-1	Soy bean	High-spin non-haem Fe(II)
Methylamine dehydrogenase	Various facultative, restricted facultative and obligate methylotrophs	None
Nitroalkane oxidase	*Fusarium oxysporum*	Noncovalently-bound FAD
Peptidylglycine amidating mono-oxygenase	Mammals	Type 2 copper
Tryptophan decarboxylase	*Catharanthus roseus*	Pyridoxal phosphate

Cu(II)-containing amine oxidases

Agmatine oxidase	*Penicillium chrysogenum*	
Benzylamine oxidase	Various yeast, mammals, fungi, birds, mussel, plants, tea leaves	
Diamine oxidase	Kidney, human placenta, plasma, other mammalian tissues, *Aspergillus niger* (yeast), fish, mussel	
Lysyl oxidase	Mammalian collagen and elastin	
Methylamine oxidase	*Candida* species (yeast) and *Arthrobacter* P1	
Phenylethylamine oxidase	*Lysophyllum aggregatum* (yeast) and *Arthrobacter globoformis*	
Plant diamine oxidase	Pea and lentil seedlings, *Euphorbia* *characias* latex, *Latyrus cicera*, *Phaseolus vulgaris*, various other plants	
Plasma amine oxidase	Pig, cow, human, horse, sheep, rabbit, rat, etc	

elastin.) It was also found that the PQQ-deficient mice accumulated less lysyl oxidase in the skin. PQQ protects against cataract damage in the lens of chicks, and protects rats against the damage of numerous hepatotoxins. Apparently, PQQ fulfills the requirement for an essential nutrient, i.e. a vitamin. Since many researchers in the field predicted that PQQ would be recognized as a vitamin, these findings were enthusiastically embraced.

NAILS IN THE COFFIN FOR PQQ AS A COVALENTLY-BOUND COFACTOR IN ANY ENZYME

In recent years, a number of enzymes were suspected of having a covalently-bound cofactor. Many of these enzymes contained essential metal ions and some contained pyridoxal phosphate. This confused the issue concerning the number and type of prosthetic groups needed for catalysis. With the discovery of PQQ, it was proposed that it could be the cofactor in such enzymes. Many of the enzymes are found in mammals, so it was not unreasonable to assume that the role of PQQ as an essential nutrient is for the efficacy of these enzymes. In the decade of the 1980's, evidence was published "proving" that the enzymes listed in Table 2 contained covalently-bound PQQ[2,5]. This was quite surprising. A number of these enzymes had been studied extensively without the slightest suspicion that they could contain this type of cofactor. Some of the enzymes certainly contain a covalently-bound quinonoid redox cofactor (e.g. the copper-containing amine oxidases and methylamine dehydrogenase).

The copper-containing amine oxidases, and in particular plasma amine oxidase, provide an excellent case study of the problems involved in unambiguously establishing the structure of a covalently-bound redox prosthetic group. Soon after the isolation of porcine plasma amine oxidase by Buffoni & Blaschko in 1964, they proposed that the plasma amine oxidase cofactor is pyridoxal phosphate[18]. In the past 25 years, a number of studies in several different laboratories using a variety of different methods claim to have confirmed this for porcine and bovine plasma amine oxidase, and porcine kidney diamine oxidase[2,5,18].

Unfortunately, accepting pyridoxal phosphate as the prosthetic group in these oxidases was not always satisfying. The results from some studies claiming pyridoxal phosphate as the cofactor did not always make quantitative sense and were often not definitive. Furthermore, this claim could not be substantiated by some groups. The u.v.-visible spectral properties of these oxidases were not consistent with pyridoxal phosphate; however, its direct interaction with the Cu(II) in these oxidases could account for this.

It was reported that [^{14}C]benzylamine was irreversibly trapped in bovine plasma amine oxidase and *Arthrobacter* P1 methylamine oxidase on reduction by $NaCNBH_3$. Moreover, when [^{3}H]$NaCNBH_3$ was used as the reductant, tritium was not trapped on these enzymes. This latter result is inconsistent with pyridoxal catalysis, but is consistent with a Schiff's base formed with a quinonoid-type cofactor[18]. The evidence supporting pyridoxal phosphate has always been circumstantial, with no direct structural data to support the contention. When PQQ came on the scene, a few investigators who were not convinced by the arguments supporting a pyridoxal derivative as the cofactor in plasma amine oxidase immediately found a solution regarding the nature of the prosthetic group.

From 1984 to 1989, no less that eight papers appeared favouring PQQ as the redox cofactor in these amine oxidases[18]. While some investigators indicated their data merely supported PQQ over pyridoxal phosphate, others flatly stated that PQQ was undeniably the cofactor. The evidence was provided by spectral analysis, chromatography, and chemical reactivities. At face value, the results were impressively in favor of PQQ as the covalently-bound cofactor in copper-containing amine oxidases.

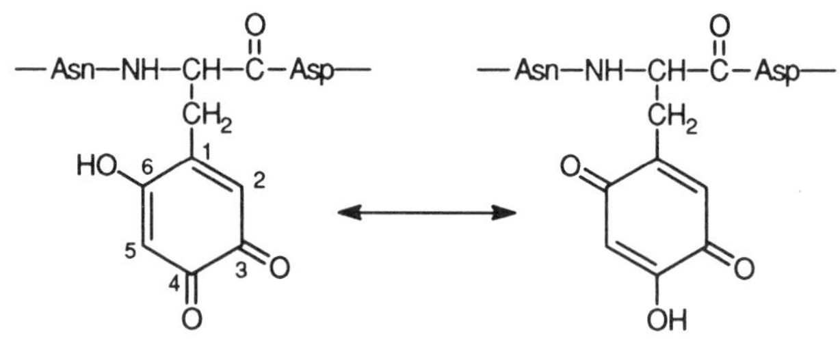

Figure 4. Structure of 6-hydroxydopa quinone (topa quinone, TQ), the cofactor for copper-containing amine oxidases

The *o*- and *p*-quinonoid tautomers of TQ are shown.

Unwittingly, the biochemists who promoted PQQ as the undeniable prosthetic group in the oxidases fell into the same trap set by pyridoxal phosphate, this time in the guise of PQQ. Again the evidence is circumstantial, in that it relied on physical and chemical properties. Fortunately, a few biochemists persisted in the belief that this type of evidence does not substitute for a direct and complete structural analysis.

The requisite research was carried out by Dr. J.P. Klinman and colleagues, and was published in 1990[19]. A phenylhydrazine-derivatized cofactor pentapeptide was isolated and analysed by a variety of approaches specifically aimed at structural elucidation. The biochemistry community was more than a little surprised by the results. The cofactor is the aminoacyl derivative of 6-hydroxydopa quinone, also known as topa quinone (TQ; Figure 4). This cofactor is a contiguous part of the polypeptide chain in the form of a modified tyrosyl residue. Recent experiments have shown that bacterial, plant, fungal and mammalian copper-containing amine oxidases all contain TQ[20].

In this day and age, when it was thought that all important mammalian cofactors had been disclosed to us, the discovery of TQ was no less than a revelation. But what about the other enzymes reported to contain covalently-bound PQQ?

About 1978, I embarked on research to discover the nature of the covalently-bound cofactor in methylamine dehydrogenase. This enzyme is found in numerous Gram-negative methylotrophic bacteria. Again, several papers have been published which provided undeniable "proof" that the methylamine dehydrogenase cofactor is PQQ[2,5]. The evidence is of the same circumstantial type provided in the amine oxidase case. While pyridoxal phosphate, flavin and pteridine derivatives were proposed, it was clear that methylamine dehydrogenase contained a quinonoid-type cofactor. The enzyme's u.v.-visible spectral properties are unlike those of any known prosthetic group. Many of the chemical and physical properties of the methylamine dehydrogenase cofactor are very reminiscent of those of PQQ, so the notion that this *o*-quinone is the cofactor of methylamine dehydrogenase is reasonable.

With several years to entrench the idea of PQQ as the cofactor of methylamine dehydrogenase in the minds of fellow biochemists, in 1986 we published a communication reporting the mass spectral analysis of a semicarbazide-derivatized cofactor peptide of methylamine dehydrogenase from bacterium W3A1. We interpreted the data within the framework of the PQQ ring system; however, some structural modifications were required. Figure 5 presents the structure deduced from the mass spectral data. While cautiously optimistic that we were on the right track, we were not satisfied that this was the correct structure. More recently, in collaboration with Dr. Dave Wemmer, we have obtained one-dimensional ^{1}H and ^{13}C n.m.r. and NOSEY and COSY

^{1}H n.m.r. data for the cofactor peptide. While the ^{1}H COSY data suggested an indole moiety as part of the cofactor, we could not piece together a structure that is consistent with all the n.m.r. and mass spectral data.

In 1989, the X-ray structure of methylamine dehydrogenase from *Thiobacillus versutus* was published by another group of investigators. In order to perpetuate the notion that PQQ is the cofactor, the "pro-PQQ" structure shown in Figure 5 was put forward to fit the electron density. It was proposed that once released from the enzyme, pro-PQQ would convert to PQQ (Figure 5).

In 1990, Drs. Lidstrom and Chistoserdov completed the sequencing of the cloned structural gene of the cofactor subunit of methylamine dehydrogenase from *M. extorquens* AM1. This established that the amino acid at both cofactor "attachment"

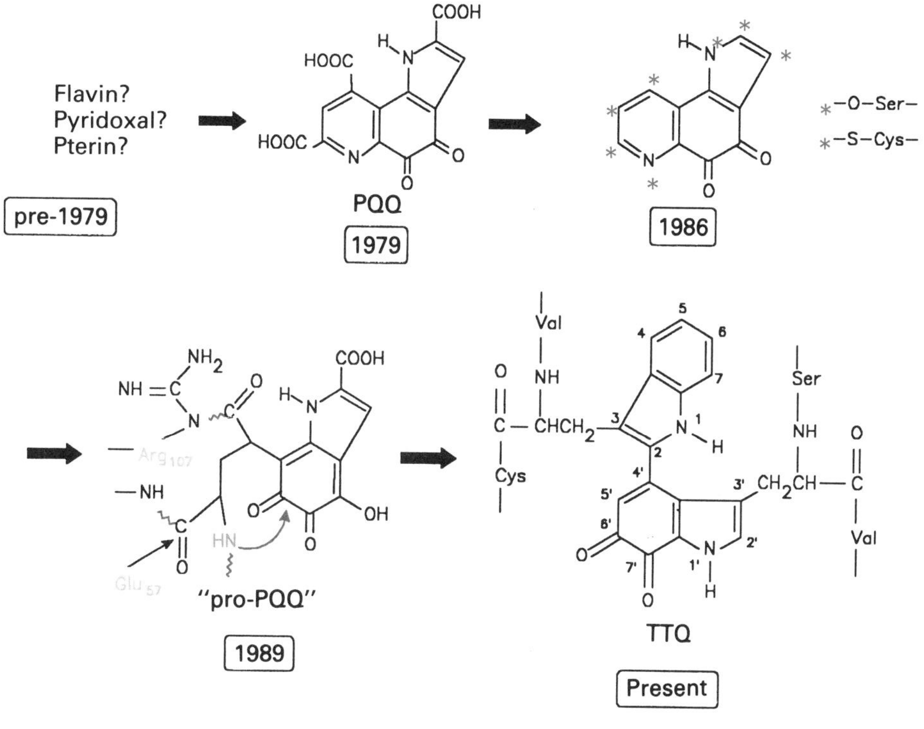

Figure 5. The transmogrification of the perceived structure of the covalently-bound cofactor of bacterial methylamine dehydrogenase
The starred positions of the 1986 structure are the postulated sites of substitution of a seryl and a cysteinyl residue for bimodal attachment of the quinone to the polypeptide chain of methylamine dehydrogenase. Unlike 2,7,9-tricarboxy-PQQ, this structure does not have any form of carboxyl groups. The "pro-PQQ" structure "deduced" from the 2.25 Å density map of methylamine dehydrogenase of *Thiobacillus versutus* was published in 1989. The dark red wriggly lines are the suggested bimodal linkages of the cofactor to the polypeptide via Glu-57 and Arg-107 (light red). According to the proposal, once these bonds are hydrolysed, the freed amino group (medium red) attacks the carbonyl carbon of the indolyl quinone (dark red curved arrow). Following this attack, dehydration and oxidation/reduction steps give 2,7,9-tricarboxy-PQQ. The present and correct bitryptophylquinone structure of the methylamine dehydrogenase cofactor is shown, including its relationship to the polypeptide chain.

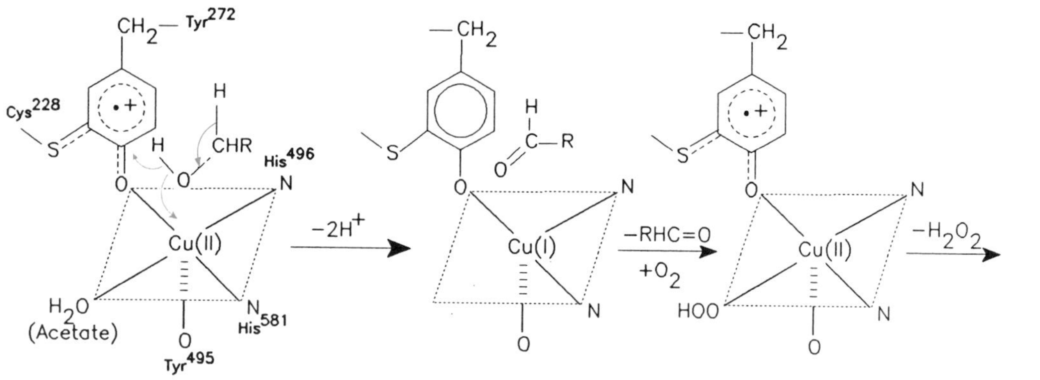

Figure 6. Structure of the cofactor site, and mechanism, of galactose oxidase
The square pyrimidal structure around Cu(II), as deduced from the 1.7 Å X-ray map of galactose oxidase, is shown at the left. The 3'-S-Cys-228–Tyr-272 cofactor (shown as the radical cation) is an equatorial ligand. Acetate is another equatorial ligand in the crystal structure. In solution, H_2O/HO- (or another anion) may replace acetate. His-496 and His-581 occupy the last two equatorial sites. Both bind through the N-3 position of their imidazole side groups. The sole axial ligand is the side group of Tyr-495. The redistribution of electrons of the aldose form of D-galactose into the Cu(II)/cofactor is displayed in dark red. After release of the product aldehyde, O_2 binds as a equatorial peroxy ligand, and is released as H_2O_2.

sites were tryptophyl residues. From our n.m.r. and mass spectral data, it was readily apparent that the cofactor structure is 2,4'-bitryptophan-6',7'-dione (tryptophan tryptophylquinone, TTQ; Figure 5)[21]. Confirmation came when this structure provided an excellent fit to the electron density in the active sites of methylamine dehydrogenase from *P. denitrificans* and *T. versutus*[22].

Galactose oxidase from *Dactylium dendroides* is another of the enzymes having an organic cofactor, in addition to essential copper. A paper appeared in 1989 incontrivertibly "proving" PQQ as the organic cofactor. In contrast, the work of Whittaker and Whittaker and coworkers provided evidence for a tyrosyl radical co-ordinated to Cu(II) in the active site of galactose oxidase. More recently, Peter Knowles and colleagues published the 1.7 Å X-ray structure of galactose oxidase which clearly shows Tyr-272 co-ordinated at its aromatic oxygen to Cu(II). Surprisingly, the phenolate ring of the tyrosyl residue is also cross-linked to Cys-228 via a thioether (Figure 6)[24,25].

Mammalian dopamine β-hydroxylase, soybean lipoxygenase-1, glutamate decarboxylase from *E. coli* and fungal laccase were all labelled as quinoproteins because all were "proven" to contain covalently-bound PQQ. Before PQQ was considered, all work on the mechanism of action of these enzymes was explained in the context of the original cofactors (see Table 2). If PQQ were indeed a second cofactor in these enzymes, the way biochemists thought about the mechanisms of these would have to be dramatically altered. As it turns out, in the past 2 years, several reports have appeared that provide overwhelming evidence that PQQ is *not* a cofactor in any of these enzymes[2,21,23].

WHAT WENT WRONG?

To date, for every putative quinoprotein that has been carried through a proper and thorough analysis, PQQ has not been found to be a covalently-bound prosthetic group. As seen in Table 2, there are other enzymes reported to contain covalently-attached PQQ. The methods employed in all cases are the same or similar to those that erroneously indicated PQQ as a cofactor in the enzymes discussed in the previous section. It is unfortunate that the data provided by these methods was offered as incontrovertible proof for the presence of PQQ. The authors of these reports are not so much at fault as is the scientific community, which so readily embraced these conclusions. As is often the case, seemingly reasonable published work soon becomes dogma. The fact that the evidence was at best circumstantial was not appreciated. The questionable methodologies measured physical and chemical properties, and should have been only viewed as preliminary screening procedures. Cases giving a positive indication of a PQQ-like species should have been followed up by a proper structural analysis. Why did the methods in question give misleading results? In most of the analyses, reverse phase high pressure liquid chromatographic (h.p.l.c.) comparisons of derivatized protein-free "cofactor" with the identical derivative of PQQ were reported. It has been pointed out that the conditions for the h.p.l.c. analysis are not very discriminating. Compounds with similar characteristics would have nearly identical retention times on reverse phase columns[19].

PQQ is biosynthesized from tyrosine and glutamate, so it is not unreasonable to speculate that a highly reactive tyrosine derivative, e.g. TQ, can condense with protein-free glutamate to produce PQQ during isolation[2]. Such a mechanism has been proposed[19]; however, the most reasonable reaction between TQ and glutamate would lead to the PQQ isomer shown in Figure 7.

It has been intimated that other protein-bound species (e.g. a tyrosyl radical) could give rise to PQQ detected during work-up and isolation. Supposedly, the precursors are formed with the participation of protein-bound metal ions, or reactive products of the enzyme reaction, e.g. O_2^- or H_2O_2. While this may be a viable hypothesis for some enzymes, it is impossible for this speculative route to explain recovery of PQQ from methylamine dehydrogenase. It is extremely difficult to envision a realistic pathway from TTQ to PQQ.

The reports of proof of covalently-bound PQQ usually included u.v.-visible and ^1H n.m.r. spectral evidence. This type of evidence is harder to refute. However, it should be pointed out that ^1H n.m.r. analysis of PQQ is not very enlightening since PQQ contains only two non-exchangable protons. Additionally, the reported n.m.r. data were often incomplete and inconclusive.

Figure 7. The structure of the most reasonable isomer of PQQ that would form by the reaction of glutamate and aminoacyl topa quinone

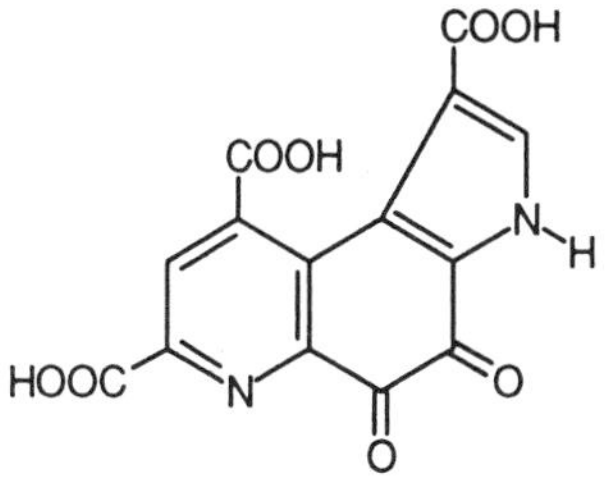

PQQ binds tenaciously to some proteins[10]. Under the right conditions it might even become covalently bound via its carboxyl group to blood and tissue proteins after absorption from the intestines.

Whichever hypothesis is chosen to explain the presence of PQQ in any enzyme listed in Table 2, it is difficult to rationalize the near-stoichiometric contents of this quinone that were routinely reported for these enzymes[21].

PROSPECTIVE

As a result of research on "quinoproteins", four new cofactors, all derived by modification of amino acids, have been discovered. The cofactors are PQQ, TTQ, TQ and the Cys-Tyr group from galactose oxidase. The latter three are certainly formed *in situ*. It has been proposed that PQQ is formed *in situ* by a process analogous to the formation of thyroxine[2,5], but more likely it is biosynthesized as are most other covalently or noncovalently-bound cofactors, that is by a pathway involving protein-free intermediates.

Reasonable pathways can be formulated for the formation of TQ from a tyrosyl residue based on reactions catalysed by known enzymes (e.g. tyrosine 3-mono-oxygenase, *p*-hydroxybenzoate hydroxylase, tyrosinase, *p*-coumarate hydroxylase, orcinol 2-mono-oxygenase). A partly or fully self-catalytic mechanism cannot be ignored. Although there is evidence for and against a direct interaction of TQ with Cu(II) in the amine oxidases, the Cu(II) and TQ are at least in close proximity[18]. Perhaps direct binding of copper to a tyrosyl residue, or copper-activated O_2 is involved in formation of TQ. A self-catalytic mechanism is more feasible for the galactose oxidase cofactor. From X-ray crystallographic data, Cu(II) is known to be co-ordinated to the phenolic oxygen of Tyr-272. After the enzyme has folded properly and Cu(II) is bound to all its aminoacyl ligands, the *ortho* position of Tyr-272 is primed for attack by the –SH/-S⁻ group of a neighboring cysteinyl residue. In contrast, biosynthesis of TTQ must occur, at least in part, with the intermediacy of an external catalyst. Methylamine dehydrogenase does not contain metal ions. It is reasonable to assume that tryptophylquinone formation precedes cross-linking. The formation of the quionoid moiety can be catalysed by any number of known enzymic processes (see the enzymes proposed above for TQ formation). The cross-linking process could be spontaneous or enzyme-catalysed[21].

Methylamine dehydrogenase is an $\alpha_2\beta_2$ periplasmic enzyme, galactose oxidase is an extracellular monomeric protein, and α_2 eukaryotic copper-amine oxidases are glycosylated and transported to their final site of action. These facts raise interesting questions concerning the sequence of events taking place during post-translation cofactor formation, membrane translocation, glycosylation, folding and subunit interactions.

Now that the exact structure of the essential organic cofactors are known in these enzymes, research aimed at elucidating the chemical mechanism of action of these enzymes can proceed on a proper course. The X-ray structures of methylamine dehydrogenase from *P. denitrificans* and *T. versutus*, and galactose oxidase from *D. dendroides*, and the manipulation of the cloned structural genes of these enzymes using modern molecular biological techniques will certainly accelerate this research. Several

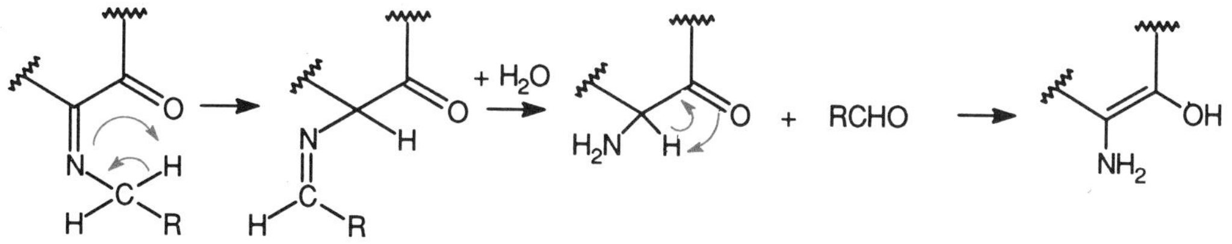

Figure 8. The proposed modified aminotransferase mechanism for plasma amine oxidase

It differs from the aminotransferase mechanism shown in Figure 3 since it involves a 1,3-prot
tropic shift (red arrows) from substrate carbon to cofactor imine carbon. Subsequent hydrolysis
and rearrangement (red arrows) produces the aldehyde product and topa aminoquinol.

laboratories are attempting to obtain X-ray crystallographic quality crystals of several TQ-containing oxidases. Hopefully, in the near future the structure of one of these enzymes will be available. The gene of a yeast copper-amine oxidase has been cloned and sequenced.

A mechanism for galactose oxidase consistent with all available observations is presented in Figure 6[24,25]. The thioether bond of Cys-228–Tyr-272 appears to have double bond character; the $C_{(\beta)}$ carbon of Cys-228 is in the plane of the Tyr-272 benzene ring. It would appear that an extended π system exists. Additionally, Trp-290 appears to have a charge transfer interaction with this π system[24].

TTQ and TQ, like PQQ, are o-quinones. Much of what is known concerning the chemistry of PQQ will be applicable to TTQ and TQ. The PQQ-based modified aminotransferase mechanism proposed for bovine plasma amine oxidase (Figure 8), remains intact with TQ. Future mechanistic work with the copper-amine oxidases will, in part, focus on the redox properties of the prosthetic group. At room temperature the Cu(II) + e$^-$ ⇔ Cu(I) transformation seems to have a similar redox potential to that of the TQH$^{\bullet -}$ + H$^+$ + e$^-$ ⇔ TQH$_2$ reaction[20].

It will be very important to precisely define the structure of the Cu(II) site and whether the metal is directly co-ordinated to TQ. The currently accepted structure of the site is shown in Figure 9. Very little is known about the reduction of O$_2$ by the reduced oxidases. It is certain that this occurs at the Cu(II)/Cu(I) site, although a copper–O$_2$ species has never been observed. Evidence exists for the formation of a superoxide intermediate[18].

Some work on the chemical mechanism of methylamine dehydrogenase has been forthcoming[10]. The observed redox states of methylamine dehydrogenase are entirely consistent with the quinonoid nature of TTQ. From the X-ray structures of the methylamine dehydrogenases of *P. denitrificans* and *T. versutus*, the dihedral angle of the planes of the indolyl rings of TTQ are ~42° and ~40°, respectively[22]. Unlike the galactose oxidase cross-linked cofactor, there is no π interaction that extends through the shared bond of aminoacyl side groups of the cofactor in methylamine dehydrogenase. This raises an important question: what is the function of the non-quinonoid indolyl moiety? An intriguing purpose for the indolyl group is as a conduit for electrons from the reduced indolylquinone to the natural electron acceptor.

Clearly, several of the mammalian proteins listed in Table 2 do not contain PQQ,

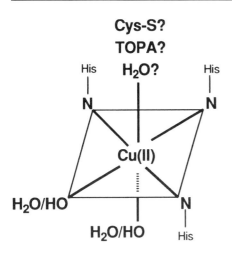

Figure 9. The Cu(II) site of copper-containing amine oxidase
Imidazole nitrogens bind three equatorial histidyl ligands. One molecule of H_2O or HO^- is an equatorial ligand and another is an axial ligand. TQ, cysteinyl or water as the sixth (axial) ligand of tetragonal bipyramidal Cu(II) has been postulated.

and indeed it is unknown whether any of the mammalian enzymes contain this cofactor. Thus, on what level does PQQ exert its nutritional effects? Possibly undiscovered mammalian PQQ-containing enzymes do exist. It may be that PQQ has a more ill-defined effect. It does seem to be present in tissue, although its detection in biological fluids and tissues has been difficult. A redox cycling assay has been developed, and was originally promoted as a method specific for PQQ, although, the method works well for detecting TQ- and TTQ-containing enzymes on denaturing electrophoresis gels. Today, the procedure is touted as one specific for any quinonoid compound in tissues and fluids, although this contention has been vigorously disputed[17,26].

It may be more correct to believe that quinonoid compounds, including PQQ, collectively have an important function in higher organisms[17,27]. The function could be to scavenge O_2^-. These quinonoid compounds may also scavenge other toxic free radicals.

CLOSING WORDS

I have used the term cofactor to describe TQ, TTQ and the Cys-Tyr group of galactose oxidase. As defined in most biochemistry textbooks, cofactor (or coenzyme) is an improper term and prosthetic group would be more correct. However, I find cofactor easier to use in written and verbal communication. Also, the existence of these new species may require a redefinition of the term cofactor. In fact, they are factors that are required in addition to those provided by the typical aminoacyl environment in proteins. In the opening paragraph of this chapter, I gave the pre-1990 definition of a quinoprotein. This definition must be altered to include enzymes containing PQQ, TQ or TTQ. These enzymes might be called PQQuinoproteins, TQuinoproteins and TTQuinoproteins.

REFERENCES

1. Salisbury, S.A., Forrest, H.S., Cruse, W.B.T. & Kennard, O. (1979) A novel coenzyme from bacterial primary alcohol dehydrogenases. *Nature (London)* **280**, 843–844

2. Duine, J.A. (1991) Quinoproteins: enzymes containing the quinonoid cofactor pyrrolo-
    quinoline quinone, topaquinone or tryptophan-tryptophan quinone.
    Eur. J. Biochem. **200**, 271–284

3. Van Spanning, R.J.M., Wansell, C.W., de Boer, T., Hazelaar, M.J., Anazawa, H., Harms, N.,
    Oltmann, L.F. & Stouthamer, A.H. (1991) Isolation and characterization of the *moxJ,
    moxG, moxI,* and *moxR* genes of *Paracoccus denitrificans*: inactivation of *moxJ, moxG,*
    and *moxR* and the resultant effect on methylotrophic growth. *J. Bacteriol.* **173**, 6948–6961

4. Tamaki, T., Fukaya, M., Takemura, H., Tayama, K., Okumura, H., Kawamura, Y., Nishiyama,
    M., Horinouchi, S. & Beppu, T. (1991) Cloning and sequencing of the gene cluster
    encoding two subunits of membrane-bound alcohol dehydrogenase from *Acetobacter
    polyoxogenes. Biochim. Biophys. Acta* **1088**, 292–300

5. Duine, J.A. (1989) PQQ and quinoprotein research – the first decade. *Biofactors* **2**, 87–94;
    Duine, J.A. & Jongejan, J.A. (1989) Quinoproteins, enzymes with pyrroloquinoline
    quinone as cofactor. *Annu. Rev. Biochem.* **58**, 403–426; Duine, J.A., Frank, J., Jzn.@
    Jongejan, J.A. (1987) Enzymology of quinoproteins. *Adv. Enzymol.* **59**, 171–212

6. Anthony, C. (1988) Quinoproteins and energy transduction, in *Bacterial Energy
    Transduction* (Anthony, C., ed.), pp. 293–316, Academic Press, New York

7. Sleath, P. R., Noar, J.B., Eberlein, G.A. & Bruice, T.C. (1985) Synthesis of 7,9-didecarboxy-
    methoxatin (4,5-dihydro-4,5-dioxo-1*H*-pyrrolo[2,3-*f*]quinoline-2-carboxylic acid) and
    comparison of its chemical properties with those of methoxatin and analogous
    o-quinones. Model studies directed towards the action of PQQ requiring bacterial oxido-
    reductases and mammalian plasma amine oxidase. *J. Am. Chem. Soc.* **107**, 3328–3338

8. Rodriguez, E.J. & Bruice, T.C. (1989) Reaction of methoxatin and 9-decarboxymethoxatin
    with benzylamine: dynamics and products. *J. Am. Chem. Soc.* **111**, 7947–7956

9. Gallop. P.M., Henson, E., Paz, M.A., Greenspan, L. & Flckiger, R. (1989) Acid-promoted
    lactonization and oxidation-reduction of pyrroloquinoline quinone (PQQ). *Biochem.
    Biophys. Res. Commun.* **163**, 755–763

10. *PQQ and Quinoproteins* (1989) (Jongejan, J.A. & Duine, J.A., eds.), Kluwer Academic
    Publishers, Dordrecht, The Netherlands

11. Ohshiro, Y. & Itoh, S. (1991) Mechanism of amine oxidation by coenzyme PQQ. *Bioinorg.
    Chem.* **19**, 169–189

12. Roderiguez, E.J., Bruice, T.C. & Edmonson, D.E. (1987) Studies on the radical species of
    9-decarboxymethoxatin. *J. Am. Chem. Soc.* **109**, 532–537

13. Houck, D.R., Hanners, J.L. & Unkefer, C.J. (1991) Biosynthesis of pyrroloquinoline
    quinone. 2. Biosynthetic assembly from glutamate and tyrosine. *J. Am. Chem. Soc.* **113**,
    3162–3166

14. Lidstrom, M.E. (1990) Genetics of carbon metabolism in methylotropic bacteria. *FEMS
    Microbiol. Rev.* **87**, 431–436; Biville, F., Turlin, E. & Gasser, F. (1989) Cloning and genetic
    analysis of six pyrroloquinoline quinone biosynthetic genes in *Methylobacteriun
    organophilum* DMS 760. *J. Gen. Microbiol.* **135**, 2917–2929

15. Turlin, E., Biville, F. & Gasser, F. (1991) Complementation of *Methylobacterium
    organophilum* mutants affected in pyrroloquinoline quinone biosynthesis genes *pqqE* and
    ppqF by cloned *Escherichia coli* chromosomal DNA. *FEMS Microbiol. Lett.* **83**, 59–64;
    Biville, F., Turlin, E. & Gasser, F. (1991) Mutants of *Escherichia coli* producing pyrrolo-
    quinoline quinone. *J. Gen. Microbiol.* **137**, 1775–1782

16. McIntire, W.S. & Weyler, W. (1988) Factors affecting the production of pyrroloquinoline
    quinone by the methylotrophic bacterium W3A1. *Appl. Environ. Microbiol.* **53**, 2183–2188;
    Ameyana, M., Matsushita, K., Shinagawa, E., Hayashi, M. & Adachi, O. (1988) Pyrrolo-
    quinoline quinone: excretion by methylotrophs and growth for microorganisms. *Biofactors*
    1, 51–53; van Kleef, M.A.G. & Duine, J.A. (1989) Factors relevant in bacterial pyrrolo-
    quinoline quinone production. *Appl. Environ. Microbiol.* **55**, 1209–1213

17. Smidt, C.R., Steinberg, F.M. & Rucker, R.B. (1991) Physiological importance of pyrrolo-quinoline quinone. *Proc. Soc. Exper. Biol. Med.* **197**, 19–26

18. McIntire, W.S. & Hartmann, C. (1992) Copper-containing amine oxidases, in *Principles and Applications of Quinoproteins* (Davidson, V.L., ed.), Marcel Dekker, New York, in the press

19. Janes, S.M., Mu, D., Wemmer, D., Smith, A.J., Kaur, S., Maltby, D., Burlingame, A.L. & Klinman, J.P. (1990) A new cofactor in eukaryotic enzymes: 6-hydroxydopa at the active site of bovine amine oxidase. *Science* **248**, 981–987

20. Brown, D.E., McGuirl, M.A., Dooley, D.M., Janes, S.M., Mu, D. & Klinman, J.P. (1991) The organic functional group in copper-containing amine oxidases. *J. Biol. Chem.* **266**, 4049–4051; Dooley, D.M., McGuirl, M.A., Brown, D.E., Turowski, P.N., McIntire, W.S. & Knowles, P.F. (1991) A Cu(I)-semiquinone state in substrate-reduced amine oxidases. *Nature (London)* **349**, 262–264

21. McIntire, W.S., Wemmer, D.E., Chistoserdov, A. & Lidstrom, M.E. (1991) A new cofactor in a prokaryotic enzyme: Trytophan trytophylquinone as the redox prosthetic group in methylamine dehydrogenase. *Science* **252**, 817–824

22. Chen, L., Mathews, F.S., Davidson, V.L., Huizinga, E.G., Villieux, F.M.D., Duine, J.A. & Hol, W.G.J. (1991) Crystallographic investigation of the tryptophan-derived cofactor in the quinoprotein methylamine dehydrogenase. *FEBS Lett.* **287**, 163–166

23. Maccarrone, M., Veldink, G.A. & Vliegenthart, J.F.G. (1991) An investigation on the quino-protein nature of some fungal and plant amine oxidases. *J. Biol. Chem.* **266**, 21014–21017

24. Ito, N., Phillips, S.E.V., Stevens, C., Ogel, Z.B., McPherson, M.J., Keen, J.N., Yadav. K.D.S. & Knowles, P.F. (1991) Novel thioether bond revealed by a 1.7 Å crystal structure of galactose oxidase. *Nature (London)* **350**, 87–90

25. Whittaker, M.M. & Whittaker, J.W. (1988) The active site of galactose oxidase. *J. Biol. Chem.* **263**, 6074–6080; Whittaker, M.M. & Whittaker, J.W. (1990) A tyrosine-derived free radical in apogalactose oxidase. *J. Biol. Chem.* **265**, 9610–9613

26. Paz, M.A., Fluckiger, R.& Gallop, P.M. (1990) Comment: Redox-cycling is a property of PQQ but not ascorbate. *FEBS Lett.* **264**, 283–284; van de Meer, R.A., Groen, B.W., Jorgenjan, J.A. & Duine, J.A. Reply: The redox-cyling assay and PQQ. *FEBS Lett.* **264**, 284

27. Gallop, P.M., Paz, M.A. & Fluckiger, R. (1990) The biological significance of quinonoid compounds in, on and out of proteins, in *Chemtracts – Biochem. Mol. Biol.* **1**, 357–37

10

Towards an understanding of C₃-C₄ photosynthesis

Stephen Rawsthorne

The Cambridge Laboratory, AFRC Institute of Plant Science Research,
John Innes Centre, Norwich NR4 7UJ, U.K.

WHAT'S IN A NAME?

Photosynthesis in higher plants occurs through a number of different mechanisms (C_3, C_4, C_3-C_4 and Crassulacean Acid Metabolism) although the most predominant amongst species is the C_3 one. In all the other mechanisms the actual net assimilation of CO_2 occurs through the same metabolic pathways as in C_3 species. Metabolic adaptations in these other photosynthetic mechanisms confer upon the plants which use them advantages in carbon and/or water economies and these have allowed them to compete and survive in environments where C_3 plants could not.

The very description C_3-C_4 intermediate photosynthesis conjures up a vision of a plant which is performing carbon dioxide assimilation by a combination of both C_3 and C_4 metabolism. As will become apparent the title given to this mechanism is not entirely helpful in describing how these plants function. However, the degree of C_4-like metabolism which is present in some intermediate species suggests a strong relationship, possibly even an evolutionary one, between this group of plants and those which utilize C_4 photosynthesis. In order to understand how C_3-C_4 intermediate plants either resemble, or differ, from those with C_3 or C_4 photosynthesis a brief description of the biochemistry behind these two well-recognized mechanisms is required. In addition, the physiological consequences of the use of these mechanisms must be considered to allow comparison of these with that of C_3-C_4 intermediate photosynthesis.

PHOTOSYNTHESIS AND PHOTORESPIRATION IN C_3 PLANTS

In higher plants which carry out C_3 photosynthesis, the photosynthetic mechanism is based on the action of ribulose-1,5-bisphosphate (RuBP) carboxylase and the regeneration of its substrate ribulose 1,5-bisphosphate by the reductive pentose phosphate pathway, or Calvin cycle, in the chloroplast. The product of CO_2 fixation by RuBP carboxylase is two molecules of 3-phosphoglycerate (a three-carbon compound; hence C_3 photosynthesis) which is either exported from the chloroplast as triose phosphate for sucrose synthesis in the cytosol or metabolized to form starch within the chloroplast. The light energy requirement of photosynthesis is for synthesis of ATP and NADPH via the electron transport pathway of the chloroplast. These compounds are then used to drive reactions of the reductive pentose phosphate pathway.

A major limitation to net CO_2 assimilation by C_3 plants occurs at the level of RuBP carboxylase because this enzyme also has an oxygenase activity in which O_2 competes with CO_2 at the active site of the enzyme. The full name of this enzyme is therefore ribulose-1,5-bisphosphate carboxylase/oxygenase. The oxygenase reaction results in the formation of 3-phosphoglycerate and phosphoglycollate (a two-carbon compound). This production of phosphoglycollate represents a drain of carbon away from the reductive pentose phosphate pathway and to recover this carbon the phosphoglycollate is metabolized through a series of reactions involving enzymes in the chloroplasts, peroxisomes, and mitochondria. In the course of this pathway two molecules of glycine are metabolized in the mitochondria to form CO_2, ammonia, and serine by the concerted actions of the enzymes glycine decarboxylase (*gdc*) and serine hydroxymethyltransferase. Serine is metabolized further to 3-phosphoglycerate and so three out of four carbon atoms are returned to the reductive pentose phosphate pathway. It is the release of CO_2 in a light-dependent manner (through provision of substrate for the oxygenase activity of RuBP carboxylase via the Calvin cycle) which leads to the term photorespiration. Not all of the CO_2 produced by decarboxylation of glycine is released and about 50% of it is recaptured by the chloroplasts before it can leave the leaf. Nevertheless, this process is not insignificant, occurring at about 30% of the rate of CO_2 assimilation in a temperate climate and so represents a major constraint on net CO_2 assimilation.

If the attached leaf of a C_3 plant is sealed into a chamber and then illuminated the CO_2 concentration within the chamber decreases until the balance point is reached between the rate of photosynthetic CO_2 fixation and the rate of photorespiratory and respiratory CO_2 release. This CO_2 concentration is termed the CO_2 compensation point (Γ) and for C_3 plants this is generally between 40 and 55 µl of CO_2/l of air depending on the exact conditions of measurement.

C_4 PHOTOSYNTHESIS

The principal difference between C_3 and C_4 photosynthesis is the introduction of a primary carboxylation reaction before net CO_2 fixation by RuBP carboxylase, which facilitates an elevation of the internal CO_2 concentration at the site of RuBP carboxylase and so prevents the competitive action of oxygen. This is only possible because of major anatomical and biochemical changes in the leaves of C_4 species. The leaves of these species are characterized by a distinctive anatomy (termed Kranz – derived

Figure 1. A bundle sheath cell of the C$_3$-C$_4$ intermediate species *Moricandia spinosa*

Notice the prominence of the mitochondria (m) on the face of the cell adjacent to the vascular tissue (v). The peroxisomes (p) and the nucleus (n) are also found in the same region of the cell and all these organelles are overlain by chloroplasts (c). Scale bar = 0.5 μm.

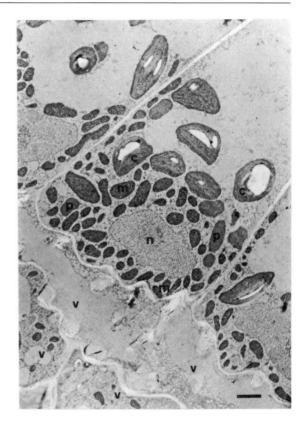

from the German for wreath) in which the vascular bundles are surrounded by large cells containing distinct chloroplasts (bundle sheath cells). These are in turn surrounded by smaller and anatomically less distinct mesophyll cells which resemble the photosynthetic mesophyll cells of a C$_3$ leaf. We now know that the mesophyll cells lack RuBP carboxylase and that the first product of CO$_2$ fixation is oxaloacetic acid (a four-carbon organic acid – hence C$_4$) formed by the carboxylation of phosphenolpyruvate in a reaction catalysed by the enzyme phosphoenolpyruvate carboxylase. This enzyme is insensitive to oxygen and is confined to the mesophyll cells. The oxaloacetate is further metabolized to malate or aspartate which are then thought to diffuse down a concentration gradient into the bundle sheath cells. These four-carbon compounds are decarboxylated directly or further metabolized and then decarboxylated by either NAD–malic enzyme, NADP–malic enzyme or phosphoenolpyruvate carboxykinase, depending on the species, which are confined to the bundle sheath cells. The three-carbon metabolite remaining after decarboxylation is then metabolized further in the bundle sheath cells before diffusing down a concentration gradient back into the mesophyll cells where phosphoenolpyruvate is regenerated to allow the cycle to continue. The CO$_2$ released in the decarboxylation reactions raises the CO$_2$ concentration in the bundle sheath cells to such an extent that the RuBP carboxylase, which is confined to these cells, does not carry out the oxygenation reaction and so photorespiration is prevented. As a consequence, the Γ values of these C$_4$ plants are generally between 0 and 5 μl of CO$_2$/l. It was during the screening of wild plants for new C$_4$ species using Γ as a means of identification that another group of plants were identified, the C$_3$-C$_4$ intermediate species.

CHARACTERIZATION OF C$_3$-C$_4$ INTERMEDIATE SPECIES

Leaf anatomy

The group name given to these species was derived originally from the intermediate nature of their Γ values between those of the C$_3$ and C$_4$ groups of plants. However, upon examination of their leaf anatomy it was clear that this also resembled that of a C$_4$ species and in many publications it is referred to as C$_4$- or Kranz-like. In all C$_3$-C$_4$ intermediate species reported to date there is an increase in the number of organelles within the bundle sheath cells relative to that in both the adjacent mesophyll cells and in bundle sheath cells of related C$_3$ species[1]. Numerous mitochondria, the peroxisomes, and many of the chloroplasts are located centripetally in the bundle sheath cells. The mitochondria are found along the cell wall immediately adjacent to the vascular cells and they are overlain by chloroplasts (Figure 1). The only exception to this is *Neurachne minor*, an Australian grass species, in which the bundle sheath cells lack a major vacuole and the abundant organelles are distributed throughout the cytoplasm[1].

CO$_2$ compensation point

As discussed above, the Γs of these species range between 30 and 8 µl of CO$_2$/l (see Table 1). This range may reflect to an extent some modulation by the leaf anatomy of the interaction between the biochemistry involved in the release and recapture of photorespired CO$_2$. For example, good negative correlations have been shown between the percentage of total leaf organelles in the bundle sheath cells and Γ in a range of C$_3$, C$_3$-C$_4$, and C$_4$ species and F$_1$ hybrids between C$_3$ and C$_3$-C$_4$ species[2,3]. The Γs of C$_3$-C$_4$ intermediate species are strongly light-dependent[1]. At photon flux densities which approach the light compensation point for photosynthesis (80–150 mol quanta m^{-2} s^{-1}) Γ can be almost that of a C$_3$ species but there is a steep decline as the light intensity increases. The Γ of C$_3$ species is unaffected by the light intensity. In addition, Γ shows a biphasic response to the O$_2$ concentration imposed during the measurement with only limited increases occurring as the O$_2$ concentration is raised to 10–20% and then a steeper C$_3$-like response beyond this range[1]. In C$_3$ species, Γ increases as a linear function of the O$_2$ concentration. The light-dependence of Γ in all the C$_3$-C$_4$ species in which this phenomenon has been examined provides evidence that the mechanism of C$_3$-C$_4$ photosynthesis is electron transport-limited (or limited by the supply of ATP and NADPH) at low CO$_2$ and is not limited by the rate of primary carboxylation by RuBP carboxylase[4].

Photorespiratory metabolism

In the leaves of plants in the genus *Moricandia* the maximum catalytic activities of a number of enzymes involved in photorespiration are similar in leaves of C$_3$ and C$_3$-C$_4$ species and in bundle sheath and mesophyll cells of the C$_3$-C$_4$ species[5]. This suggests that the capacities of photosynthetic cells in these leaves for photorespiratory metabolism are similar. The critical difference between C$_3$ and C$_3$-C$_4$ species is that

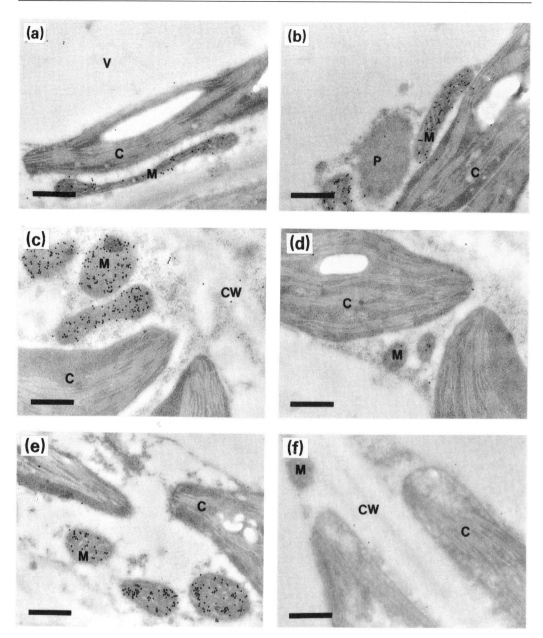

Figure 2. Immunogold labelling of mitochondria in leaf cells of representative C$_3$ and C$_3$-C$_4$ intermediate species of *Moricandia* by antibodies to the P subunit of glycine decarboxylase

M. foleyi (C$_3$) (a) bundle sheath, (b) mesophyll; *M. arvensis* (C$_3$-C$_4$) (c) bundle sheath, (d) mesophyll; *M. nitens* (C$_3$-C$_4$) (e) bundle sheath, (f) mesophyll. Scale bar = 0.5 μm. C, chloroplast; CW, cell wall; M, mitochondria; P, peroxisome; V, vacuole.

gdc is confined to the bundle sheath cells of the C$_3$-C$_4$ intermediate species whereas it is present in the mitochondria of all photosynthetic cells in C$_3$ species[5]. This was first demonstrated in the genus *Moricandia* by using immunogold labelling for the P subunit of the *gdc* complex on embedded leaf tissue (Figure 2). This important immunocytochemical observation has been confirmed by measuring enrichment of *gdc* activity in protoplast fractions derived from bundle sheath cells relative to that in protoplasts from mesophyll cells of the C3-C4 intermediate species *Moricandia arvensis*[6]. The same differential localization of *gdc* was subsequently demonstrated in leaves of a wide range of C3-C4 intermediate species from a number of different genera including *Flaveria, Mollugo and Panicum*[7]. Serine hydroxymethyltransferase activity is also enriched in bundle sheath cells of *M. arvensis*[6].

The absence of the P subunit from the mesophyll prevents the decarboxylation of the glycine produced as a consequence of the oxygenase reaction of RuBP carboxylase in these cells. Whilst this prevents release of photorespiratory CO_2 in the mesophyll cells, the carbon in the glycine which is drained away from the reductive pentose phosphate pathway in the mesophyll must be returned to prevent this regenerative cycle from running down. This implies that the glycine must move from the mesophyll cells to the bundle sheath cells to be decarboxylated. Carbon must then return from the bundle sheath cells to the reductive pentose phosphate pathway in the chloroplasts of the mesophyll cells to allow continued CO_2 fixation.

We have proposed a scheme for photorespiratory metabolism in the leaf of *M. arvensis* (Figure 3) which is likely to be equally applicable to all C3-C4 intermediate

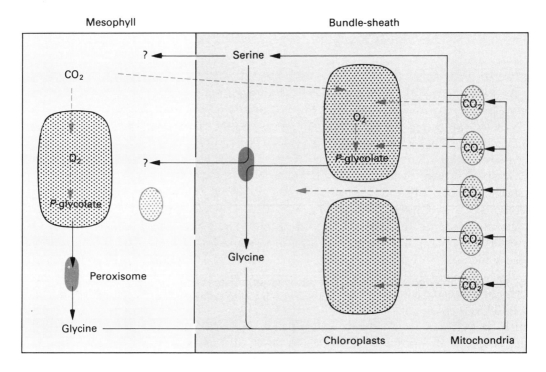

Figure 3. A proposed pathway for photorespiration in the leaves of C3-C4 intermediate species (after[5])
Movements of metabolites or gases are indicated by solid or broken lines, respectively.

species including those which are reported to have some elements of C_4 photosynthesis (see below). This scheme implies that the major if not the sole site of release of photorespiratory CO_2 is in the bundle sheath cells of most if not all C_3-C_4 intermediate species. Because these mitochondria are in close association with the chloroplasts it is thought that this must greatly enhance the potential for recapture of the photorespired CO_2 in these species[5]. Indeed, the recapture of photorespired CO_2 in the leaves of *M. arvensis* (C_3-C_4) is 50% greater than that achieved by the related C_3 species *M. moricandioides*[8].

A second implication from this scheme is that there must be movement of the glycine produced in the mesophyll to the bundle sheath cells. There is no definitive evidence that demonstrates how this is likely to occur but it is tempting to draw an analogy to the diffusion-based mechanism used in C_4 plants[9], which facilitates the operation of the C_4 cycle, by using a number of lines of supportive evidence. First, the number of plasmodesmata which interconnect the bundle sheath and mesophyll cells are similar in C_3-C_4 and C_4 *Panicum* species[10]. The capacity for metabolite exchange between the bundle sheath and mesophyll of these C_3-C_4 *Panicum* species is therefore much greater than that of related C_3 species. Second, the concentrations of glycine and serine in leaves of a C_3-C_4 intermediate *Moricandia* are two-fold greater on a chlorophyll basis than those in leaves of a C_3 *Moricandia* and are four-fold greater on a leaf area basis than the concentrations of metabolites involved in cell to cell movement in C_4 plants[11]. Third, these metabolite pools change in response to environmental perturbation in a manner entirely consistent with maintenance of a diffusion gradient between the cell types[11].

C_3-C_4 METABOLISM IN THE GENUS *Flaveria*

The discovery of C_3-C_4 plants with intermediate CO_2 compensation points prompted a number of publications which attempted to show that this change in physiology was due to enhanced activities of enzymes of the C_4 pathway. However, except for species in the genus *Flaveria* it has subsequently been shown that this is not the case. For example, following photosynthetic fixation of $^{14}CO_2$ by a C_3-C_4 intermediate *Moricandia* species there is no movement of ^{14}C from C_4 acids into intermediates of the reductive pentose phosphate pathway[8]. This is the most critical demonstration of C_4 metabolism and for C_3-C_4 intermediate species in the genus *Flaveria* there is undoubtedly evidence for significant fixation of $^{14}CO_2$ into C_4 acids during photosynthesis. Within the genus there is a more-or-less continuous range of plants between those which are true C_3 or C_4 species[12,13,23]. This range includes C_3-C_4 intermediate species which barely fix CO_2 into C_4 acids and must therefore rely on the recapture of photorespired CO_2 to reduce Γ[12]. Other C_3-C_4 intermediates in the genus incorporate up to 50% of $^{14}CO_2$ into C_4 acids[12] while "C_4-like" has been used to describe species which fix between 80 and 95% of $^{14}CO_2$ into C_4 acids[13,14]. In the C_3-C_4 intermediate *Flaveria* species the incorporated ^{14}C label does not pass exclusively into the reductive pentose phosphate pathway during a pulse-chase experiment but is found in glycine and serine (up to 40% of the initial label incorporated) and rather unusually fumarate (up to 35% of the initial label)[12]. In the C_3 *Flaveria* species used in the same study only small amounts of label were detected in glycine and serine, and fumarate. The patterns of ^{14}C-labelling in the C$_3$-C$_4$ species are clearly unlike that of a true C$_4$ species in which the ^{14}C label passes directly from the C$_4$ acids into reductive pentose phosphate pathway intermediates, then into sucrose and does not appear in glycine, serine, or fumarate[12]. The "C$_4$-like" species fix so little CO$_2$ directly by RuBP carboxylase that their metabolism very closely resembles that of a true C$_4$ species[13,14].

The accumulation of ^{14}C label into fumarate in some *Flaveria* species may be explained in part by the large pools of this metabolite in the leaves of these particular species[4]. The appearance of significant amounts of label in fumarate may not therefore represent an end-point of metabolism but the dilution of the radioactively-labelled fumarate into a large *in vivo* pool. A similar explanation may account for the accumulation of label in glycine and serine if the glycine pool in C$_3$-C$_4$ intermediate *Flaveria* is larger than that of the C$_3$ species as is seen in *Moricandia*[11].

The [14]CO$_2$ labelling patterns seen in this range of *Flaveria* species are reflected broadly in both the activities of enzymes of the C$_4$-cycle in the leaves and in the distribution of these enzymes between the bundle sheath and mesophyll cells. Clearly, without partitioning of the carboxylating and decarboxylating elements of C$_4$ metabolism between the appropriate cell types, futile cycling of CO$_2$ would occur within those cells. Only in the C$_4$-like species do the activities of C$_4$ cycle enzymes show the intercellular distribution and activities which would allow a functional C$_4$ pathway to occur[13,14].

Clearly, aspects of their photosynthetic metabolism differentiate the C$_3$-C$_4$ intermediate species of *Flaveria* from other C$_3$-C$_4$ species. However they still have the same differential distribution of *gdc* between their bundle sheath and mesophyll cells as other intermediate species and this could explain their lower Γ values. More work is required to allow us to understand what roles the CO$_2$ fixation into C$_4$ acids and the metabolism of fumarate play in photosynthetic metabolism of C$_3$-C$_4$ intermediate *Flaveria* species.

REGULATION OF EXPRESSION OF GENES ENCODING GLYCINE DECARBOXYLASE

Glycine decarboxylase is a multisubunit enzyme complex in bacteria, plants and animals and comprises P (pyridoxal phosphate-containing), H (lipoic acid-containing), T (transferase) and L (lipoamide dehydrogenase) subunits[15,16]. The *gdc* complex when isolated from plant mitochondria at a high protein concentration is stable and contains approximately 2 P-protein dimers : 27 H-protein monomers : 9 T-protein monomers : 1 L-protein dimer[16]. At lower protein concentrations the subunit proteins tend to dissociate from the complex and decrease in activity[16]. It is probable that the enzyme exists as a stable complex in the matrix of plant mitochondria where the protein concentration is estimated to be as high as 130 mg/ml, one-third of which is accounted for by the *gdc* complex[16].

Following the discovery of a differential distribution of the P subunit of *gdc* and of the activity of this enzyme in leaves of intermediate plants it has become clear that we need to understand how this process is regulated and whether this regulation is co-ordinated for all the subunits of the complex. The cloning of genes encoding the subunits of *gdc* is currently an active area in both plant and mammalian areas (because of the role of *gdc* mutations in hyperglycinaemia). Sequences of cDNAs

Table 1. Phylogenetic distribution of C_3-C_4 photosynthesis
The occurrence of C_4 photosynthesis in the family and/or genus is given (*), along with the CO_2 compensation points (Γ) of representative species. The data are drawn from [1] except for (a)[23], (b)[8], (c)[7] and (d)[24].

Family	Genus	Species	Γ	Photosynthetic type
Dicotyledonae				
Asteracae*	Flaveria	F. pringlei	62[a]	C_3
		F. linearis	27[a]	C_3-C_4
		F. floridana	10[a]	C_3-C_4
		F. brownii	6[a]	C_4-like
		F. trinervia	4[a]	C_4
	Parthenium	P. hysterophorus	22	C_3-C_4
Amaranthaceae	Alternathera*	A. ficoides	22	C_3-C_4
		A. tenella	18	C_3-C_4
Aizoaceae	Mollugo*	M. verticillata	30	C_3-C_4
Brassicaceae	Moricandia	M. moricandioides	42[b]	C_3
		M. arvensis	16[b]	C_3-C_4
		M. nitens	9[c]	C_3-C_4
Monocotyledoneae				
Cyperaceae	Eleocharis	E. pusilla	30[d]	C_3-C_4
Poaceae	Neurachne	N. minor	5	C_3-C_4
	Panicum	P. milioides	15	C_3-C_4

have been published within the last 2 years for the P, H, T and L subunits of pea (*Pisum sativum*) and of the human, chicken, and bovine liver *gdc* (see [17,18] and references therein). In *P. sativum* these genes are present as one or two copies per haploid genome. In my own laboratory we have been gaining an understanding of the temporal co-ordination and organ specificity of expression of these genes in *P. sativum*[17,18]. The genes are all induced by light in leaves and show a broadly similar pattern of developmental expression but they are not expressed uniformly in all plant organs. For example, Northern blot analysis shows H subunit mRNA to be absent from roots whereas P subunit mRNA is readily detected[17]. The antibody and cDNA probes are now being used as tools to study how *gdc* differential expression is achieved in leaves of C_3-C_4 intermediate plant species. Recent unpublished work using immunogold labelling and *in situ* hybridization has revealed that *only* the P subunit is absent from the mesophyll mitochondria of *M. arvensis* and that this appears to be due to cell-specific transcriptional regulation of the P gene.

DISTRIBUTION OF C_3-C_4 INTERMEDIATE SPECIES

The C_3-C_4 species are generally found in arid or semi-arid regions of the world[1]. *Moricandia* species, for example, are found in Southern Mediterranean and North African regions. Whilst it is tempting to suggest that the C_3-C_4 mechanism must provide the plant with an adaptive advantage in dry areas, it must be remembered that the intermediate species are found in the same habitats as their C_3 relatives. Similar observations have been made for the distributions of C_4 and related C_3 species even though the C_4 mechanism clearly leads to a considerable improvement in the water use efficiency (amount of dry matter gained per unit of water lost from the plant through transpirational processes). Clearly, factors other than the photosynthetic mechanism used by a plant contribute to its adaptation to a particular environment. Perhaps of more interest is the phylogenetic distribution of C_3-C_4 species in the higher plant kingdom. This mechanism has occurred independently in evolutionarily unrelated species with remarkable conservation of the basic elements of the character (anatomical changes and differential distribution of *gdc*). There are eight genera which contain C_3-C_4 species in six plant families (Table 1). Of these, only two genera, *Moricandia* and *Parthenium*, do not contain species which have the C_4 mechanism. *Moricandia* is of interest since it is the only Crucifer which has been reported to have a photosynthetic mechanism other than C_3. This prompts the question as to whether these species represent a step in the evolution of the C_4 mechanism from the older C_3 one.

C₃-C₄ INTERMEDIATES AND THE EVOLUTION OF C₄ PHOTOSYNTHESIS

There has been much speculation as to the significance of C_3-C_4 intermediates in the evolutionary transition from a C_3 to a C_4 species. The water use efficiency of the intermediate species has been studied with this question in mind, and while not all experiments are conclusive there is an indication that improvements in this parameter occur relative to C_3 species under certain environmental conditions[19], perhaps most significantly under conditions of low CO_2 concentration[20,21] (about 50% of the concentration in today's atmosphere). It is possible that the evolution of C_4 photosynthesis began when the CO_2 concentration of the atmosphere was much less than the present-day level. Because under those conditions the oxygenation reaction of RuBP carboxylase would be more prominent than today, there would be strong selection pressure for mechanisms which more efficiently recovered the CO_2 lost in photorespiration or better still prevented it. The evolution of C_4 photosynthesis may therefore have been driven by photorespiratory pressures which in turn led to improvements in the water use efficiency. The similar anatomical development of bundle sheath cells and the consistent absence of *gdc* from mesophyll cells in all the C_3-C_4 species studied, combined with a potential for improved water use efficiency, indicates that the differential distribution of *gdc* could have been a primary event in the evolution of C_4 photosynthesis. Our recent work[22] on the cell-specific distribution of *gdc* subunits in a range of C_3-C_4 species from different genera has shown that where the genus contains C_4 species all of the *gdc* subunits are absent from the mesophyll cells and this is also true for the C_4 species within those genera. One could therefore speculate that the absence of only the P subunit, as in *M. arvensis*, was an initial event in the evolution of C_4 photosynthesis and that loss of the other subunits has occurred in those C_3-C_4 species in genera which have had sufficient time to allow evolution of fully developed C_4 photosynthesis. Limited, or partial, C_4 cycle activity in C_3-C_4 intermediate *Flaveria* species may then represent a stage in C_4 evolution which is more advanced towards a functional C_4-syndrome than the stage represented by C_3-C_4 Panicums and Moricandias.

REFERENCES

1. Edwards, G.E. & Ku, M.S.B. (1987) Biochemistry of C_3-C_4 intermediates, in *The Biochemistry of Plants*, volume 10, pp. 275–325, Academic Press, London

2. Brown, R.H., Bouton, J.H., Evans, P.T., Malter, H.E. & Rigsby, L. (1984) Photosynthesis, morphology, leaf anatomy and cytogenetics of hybrids between C_3 and C_3/C_4 *Panicum* species. *Plant Physiol.* **77**, 653–658

3. Brown, R.H. & Hattersley, P.W. (1989) Leaf anatomy of C_3-C_4 species as related to evolution of C_4 photosynthesis. *Plant Physiol.* **91**, 1543–1550

4. Rawsthorne, S., von Caemmerer, S., Brooks, A. & Leegood, R.C. (1991) Metabolic interactions in leaves of C_3-C_4 intermediate plants, in *Metabolic Interactions of Organelles in Photosynthetic Tissues*, S.E.B. Seminar series, vol. 45 (Tobin, A.K., ed.), Cambridge University Press, Cambridge (in the press)

5. Rawsthorne, S., Hylton, C.M., Smith, A.M. & Woolhouse, H.W. (1988) Photorespiratory metabolism and immunogold localization of photorespiratory enzymes in leaves of C_3 and C_3-C_4 intermediate species of *Moricandia*. *Planta* **173**, 298–308

6. Rawsthorne, S., Hylton, C.M., Smith, A.M. & Woolhouse, H.W. (1988) Distribution of photorespiratory enzymes between bundle-sheath and mesophyll cells in leaves of the C_3-C_4 intermediate species *Moricandia arvensis* (L.) DC. *Planta* **176**, 527–532

7. Hylton, C.M., Rawsthorne, S., Smith, A.M., Jones, D.A. & Woolhouse, H.W. (1988) Glycine decarboxylase is confined to the bundle-sheath cells of leaves of C_3-C_4 intermediate species. *Planta* **175**, 452–459

8. Hunt, S., Smith, A.M. & Woolhouse, H.W. (1987) Evidence for a light-dependent system for reassimilation of photorespiratory CO_2, which does not include a C_4 cycle, in the C_3-C_4 intermediate species *Moricandia arvensis*. *Planta* **171**, 227–234

9. Leegood, R.C. & Osmond, C.B. (1990) The flux of metabolites in C_4 and CAM plants, in *Plant Physiology, Biochemistry, and Molecular Biology* (Dennis, D.T. & Turpin, D.H., eds.), pp. 274–298, Longman, London

10. Brown, R.H., Bouton, J.H., Rigsby, L. & Rigler, M. (1983) Photosynthesis of grass species differing in carbon dioxide fixation pathways. VIII. Ultrastructural characteristics of *Panicum* species in the *Laxa* group. *Plant Physiol.* **71**, 425–431

11. Rawsthorne, S. & Hylton, C.M. (1991) The relationship between the post-illumination CO_2 burst and glycine metabolism in leaves of C_3 and C_3-C_4 intermediate species of *Moricandia*. *Planta* **11**, 122–126

12. Monson, R.K., Moore, B.d., Ku, M.S.B. & Edwards G.E. (1986) Co-function of C_3- and C_4-photosynthetic pathways in C_3, C_4, and C_3-C_4 intermediate *Flaveria* species. *Planta* **168**, 493–502

13. Moore, B.d., Ku, M.S.B. & Edwards, G.E. (1989) Expression of C_4-like photosynthesis in several species of *Flaveria*. *Plant Cell Environ.* **12**, 541–549

14. Cheng, S.-H., Moore, B.d., Wu, J., Ku, M.S.B. & Edwards, G.E. (1989) Photosynthetic plasticity in *Flaveria brownii*. Growth irradiance and the expression of C_4 photosynthesis. *Plant Physiol.* **89**, 1129–1135

15. Kikuchi, G. (1973) The glycine cleavage system: composition, reaction mechanism, and physiological significance. *Mol. Cell. Biochem.* **1**, 169–187

16. Oliver, D.J., Neuburger, M., Bourguignon, J. & Dounce, R. (1990) Interaction between the component enzymes of the glycine decarboxylase multienzyme complex. *Plant Physiol.* **94**, 833–839

17. Turner, S.R., Ireland, R.J. & Rawsthorne, S. (1992) Cloning and characterization of the P subunit of glycine decarboxylase of pea (*Pisum sativum*). *J. Biol. Chem.* **267**, 5355–5360

18. Turner, S.R., Ireland, R.J. & Rawsthorne, S. (1992) Cloning and characterization of a cDNA encoding the L subunit of glycine decarboxylase of pea (*Pisum sativum*). *J. Biol. Chem.*, in the press

19. McVetty, P., Austin, R.B. & Morgan, C.L. (1989) A comparison of the growth, photosynthesis, stomatal conductance and water use efficiency of *Moricandia* and *Brassica* species. *Ann. Botany* **64**, 87–94

20. Brown, R.H. & Simmons, R.E. (1979) Photosynthesis of grass species differing in CO_2 fixation pathways. I. Water-use-efficiency. *Crop Sci.* **19**, 375–379

21. Monson, R.K. (1989) The relative contributions of reduced photorespiration, and improved water- and nitrogen-use efficiencies, to the advantages of C₃-C₄ photosynthesis in *Flaveria*. *Oecologia* **80**, 215–221

22. Morgan, C.L., Turner, S.R. & Rawsthorne, S. (1992) Cell-specific distribution of glycine decarboxylase in leaves of C₃, C₄ and C₃-C₄ intermediate species, in *Molecular, Biochemical and Physiological Aspects of Plant Respiration* (Lambers, H. & van der Plas, L., eds.), SPB Publications, The Hague, in the press

23. Ku. M.S.B., Wu, J., Dai, Z., Scott, R.A., Chu, C. & Edwards, G.E. (1991) Photosynthetic and photorespiratory characteristics of *Flaveria* species. *Plant Physiol.* **96**, 518–528

24. Bruhl, J.J., Stone, N.E. & Hattersley, P.W. (1987) C₄ acid decarboxylase enzymes and anatomy in sedges (Cyperaceae): first record of NAD-malic enzyme species. *Aust. J. Plant Physiol.* **14**, 719–728

11

Biosensors: principles and practice

Jeffrey D. Newman and Anthony P.F. Turner

*Biotechnology Centre, Cranfield Institute of Technology, Cranfield,
Bedford MK43 0AL, U.K.*

INTRODUCTION

Biosensors have been proffered as solutions to multifarious analytical problems in clinical chemistry, food analysis, biotechnology, environmental monitoring and defence. The enormous breadth of proposed applications is matched only by the variety of permutations of transducer and biological components that have been described in the literature. Selected examples include home monitoring of glucose by diabetics, rapid detection of microbial contamination in food and the measurement of ethanol for fermentation process control. The relative activity in different areas of biosensor technology (amperometric, potentiometric, optical and others) is illustrated in Figure 1.

Figure 1. Biosensor publications in 1990
Solid boxes indicate papers; hatched boxes indicate patents.

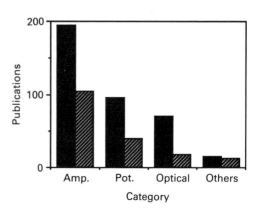

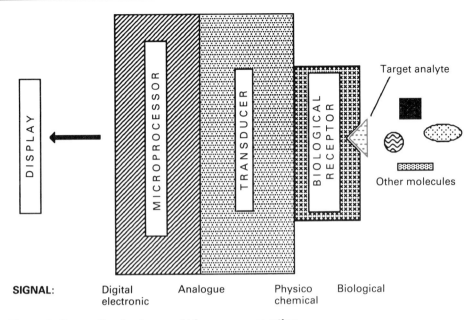

SIGNAL: Digital   Analogue   Physico   Biological
                 electronic   chemical

Figure 2. Generalized scheme of biosensor operation

Early success[1] with electrochemical devices together with their relative simplicity and flexibility accounts for the dominance of research activity in this area. Optical technology, however, has advanced rapidly in recent years as a result of activity in the telecommunications industry and offers interesting opportunities, particularly in combination with biological affinity reactions. The considerable investment of time and money worldwide in the field of biosensors over the last decade is beginning to crystallize in the form of novel commercial devices for analysis.

This essay briefly reviews the essential elements of biosensor technology and illustrates a selection of the most important developments that have led, or are leading, to practical devices. Biosensors have been the subject of several books recently[2-5], in which a thorough documentation of the subject area can be obtained. In this text, the components that make up biosensors will be described and some selected examples will be given that illustrate important principles and advances.

BIOLOGICAL COMPONENTS

One of the chief attractions of biosensors is the remarkable specificity that biological molecules confer on them. Biological components can be used to make either catalytic or affinity sensors. In both cases, it is binding reactions that provide the specificity, but with catalytic sensors it is a change in concentration of a component due to the catalysed reaction that is detected, whereas with an affinity sensor it is the binding event itself that is monitored.

Enzymes are by far the most commonly used components in catalytic biosensors, but microbial, plant or animal cells, or subcellular organelles, have been used where the use of enzymes has proved impractical or impossible. Many of the known enzymes are now available commercially, and advances in enzyme technology should enhance biosensor development by increasing enzyme availability. Recently, there has been a rapid increase of interest in the operation of enzymes in non-aqueous media. When used in the organic phase, enzymes have been shown, not only to retain activity, but to offer certain features and benefits. Improved thermal stability and altered substrate specificity, together with the possibility of detecting previously inaccessible analytes, is of interest for biosensor applications. This novel area has been reviewed recently[16].

Most affinity sensors reported to date are based on immunological principles, but other receptor molecules can be used. Lectins, plant proteins that bind specific carbohydrates, can be used in biosensors and cell receptors have been used in gas sensing. Also, bioreceptors, synthetic molecules with affinity properties, can be applied in biosensors. Devices need not be restricted to the use of a single biological component. Various combinations, such as more than one enzyme or an antibody and an enzyme can be co-immobilized to provide new or improved analytical capabilities.

The component parts of a biosensor must be held in close proximity to one another to facilitate efficient operation. The localization of biological components at or close to the transducer surface permits continuous or repeated use, maximizes the response by concentrating the biocomponent and often increases stability, particularly in enzymes. The methods used, collectively known as immobilization techniques, can be separated into: entrapment with a membrane, gel or microcapsule, physical or chemical adsorption, crosslinking between molecules and covalent bonding to insoluble supports. The area has been covered comprehensively in a recent article[17].

TRANSDUCERS

In a biosensor, a transducer is used to convert the biological event, be it a catalytic response or a binding reaction, into a form where it can be read directly or further processed by a microprocessor. The mode of operation of a generalized biosensor is outlined in Figure 2. This shows that a biological event is detected as a physical or chemical change by the transducer which converts this into an analogue signal that can be digitized by a microprocessor and displayed.

Since the transducer is in intimate contact with the biological component and the sample medium, it must be compatible with the sample and amenable to an appropriate immobilization procedure. A wide range of transducers may be used in biosensors and these can be broadly divided into electrochemical, optical, piezoelectric and calorimetric devices. Examples of these are outlined in Table 1, and are discussed more fully in the following sections.

Table 1. Examples of transducer types

Transducer	Examples
Electrochemical	Clark oxygen electrode, mediated electrodes, ion-selective electrodes, chemical field-effect transistors, light-addressable potentiometric sensors
Optical	Photodiodes, waveguide systems, integrated optical sensors
Piezoelectric	Quartz crystals, surface acoustic wave devices
Calorimetric	Thermistor, thermopile

Electrochemical devices

Many biological systems consume and/or produce electroactive species which can be monitored by electrochemical techniques such as amperometry, potentiometry or by conductimetric methods.

Amperometry has been used for the measurement of a wide variety of species in many applications, not least in biological media. Under suitable conditions, it is possible to measure concentrations in the nanomolar range directly, with a dynamic range of up to four orders of magnitude or more. The technique classically utilizes a three-electrode system in which a potential is applied between a working electrode and a reference electrode by the use of a counter or auxiliary electrode. In this way, effectively zero current is passed through the reference electrode so that the reference potential is not perturbed. The resulting current measured is due to the oxidation or reduction of the species of interest at the working electrode. The current is directly proportional to the concentration of the analyte at the working electrode surface and is therefore dependent on the rate of mass transfer of this component to the surface and to the charge transfer kinetics.

For many practical devices, it is more convenient to combine the counter and reference electrodes. Since a current is passed, this causes a drift in the reference potential, but for short periods of operation this may not be a problem. Amperometric devices may be subdivided into cathodic (reducing) or anodic (oxidizing) sensors. Both types have been extensively used in biosensors. The classical working electrode material is mercury. For biosensors however, solid electrodes based on platinum, carbon and gold have been preferred. Finding a suitable method for surface preparation and the occurrence of fouling during use, especially in biological samples, are particular problems with solid electrodes. The Clark electrode[1], incorporating an oxygen-permeable polymer membrane that excludes interferents, overcomes many of these difficulties and is still commonly used, particularly in research applications.

A related constant-potential technique is coulometry, which involves the measurement of the total charge transfer rather than the current. This technique follows Faraday's law[6]:

$$N = Q/nF$$

where N = number of moles of reactant consumed, Q = total charge passed (coulombs), n = number of electrons involved in the reaction and F = Faraday constant (96 487 coulombs/mole).

It is therefore an absolute method, independent of the electrode and reaction kinetics. The technique is usually carried out with small sample volumes to reduce the time required to consume all of the electroactive species.

A potentiometric sensor operates under conditions of near-zero current flow and measures the difference in potential between the sensing element and a reference electrode. The response of these devices is usually proportional to the logarithm of the activity of an ion in solution, characterized by the Nernst equation[6]:

$$E_{ISE} = E_0 + 2.303 \, (RT/zF) \log a_i$$

where E_0 = a constant, independent of a_i, R = gas constant (8.314 J/K), T = temperature (K), z = charge number of ion i and F = Faraday constant.

This corresponds to a linear relationship between the measured potential, E_{ISE} and the logarithm of the activity of the ion i (a_i) with a slope of 59.16 mV for each 10-fold change of activity of a monovalent ion at 25 °C.

Ion-selective electrodes are the most important of this class of transducer, with many applications in biosensors. Specificity is conferred by selective membranes which may be formed from metal salts or polymer membranes containing ion-exchangers or neutral carriers. The construction and operation of these ion-selective electrodes are discussed in detail elsewhere[8].

Miniaturized potentiometric devices can be fabricated by deploying field effect transistors as transducers. This approach was discovered in 1970 by Bergveld[9] who, instead of connecting an ion-selective electrode to an impedance-converting field effect transistor, used the transistor directly as an ion-selective electrode. Deposition of a wide range of permselective membranes on the silicon nitride gate layer of the field effect transistor enables a wide variety of ions to be detected. The small dimensions and ready availability of microelectronics fabrication technology makes the field effect transistor an attractive transducer for biosensor applications. However, problems in depositing and maintaining biological molecules on devices of this type have limited their success. A possible area where developments are possible is in the production of multi-sensors, in which a single device may contain many sensing elements.

Light-addressable potentiometric sensors are based on a silicon wafer possessing a chemically modified surface. A measurement is made by illuminating the back of the wafer with a light-emitting diode which causes a photocurrent to flow. Modulation of the light results in an alternating current which is pH-dependent. A chemical reaction changing the pH at the sensor surface alters the current. If an external potential is applied to maintain a constant current, the reaction rate can be monitored by the change in this potential.

Suggested further reading: Bard & Faulkner (1980)[6], Wang (1988)[7].

Optical devices

Emission and absorption of electromagnetic radiation are useful techniques that have been used in chemical analysis for over 100 years. Emission spectroscopy, atomic absorption, and Raman scattering are some examples of techniques that have been used. Modern advances in miniaturization make possible the incorporation of optical analysis into small, cheap and simple-to-operate sensors.

The change in the light absorbance characteristics of a reagent layer on interaction with an analyte is the simplest form of optical detection. Optical fibres are commonly used to carry light to and from a reaction area. Suitable materials for the construction of optical fibres, which transmit light by multiple internal reflections, include glasses and plastics. Indicators immobilized on optical sensors can be used to react reversibly with an analyte to form a product with different optical properties to the original reagent. This principle has been used for many analytes and is typified by an optical pH sensor in which an indicator in its base form is protonated by the analyte, the hydrogen ion. These sensors have a working range of 1–2 pH units around the pK_a of the indicator. Considerable research has been focussed on indicators suitable for use in the physiological pH range, but little has been done outside this range. Most oxygen sensors based on optical principles utilize fluorescence quenching. The fluorescent material can be incorporated onto the sensor in the solid phase, or it can be in solution separated from the sample by an hydrophobic oxygen-permeable membrane. The measurement can also be based on a change of intensity or a shift in the spectrum.

A light beam emanating from a medium of higher refractive index (n_1) striking the interface with a medium of lower refractive index (n_2) undergoes total internal reflection if the angle of reflection is larger than the critical angle (ϕ_c):

$$\phi_c = \sin^{-1}(n_2/n_1)$$

When light is reflected in an optical device, there is a decay of energy away from the point of reflection into an external medium of a different refractive index. This electromagnetic phenomenon is known as the evanescent wave and can be used to detect a change in refractive index in close proximity to the optical interface. There are three principal techniques, commonly used in optical biosensors, that utilize this effect. Attenuated total reflection relies on the absorption of light of specific wavelengths by molecules at the waveguide surface. This absorption is proportional to the concentration of these molecules present. Total internal reflection fluorescence occurs when the evanescent wave is used to excite fluorescent molecules bound to the waveguide surface. It is argued that a washing step, as required in conventional immunoassays to separate bound from unbound antibody, is no longer required with this technique since fluorescent molecules in the bulk solution are not excited. When a glass substrate is coated with a thin metal layer, it is possible to excite an electromagnetic wave (a surface plasmon) which propagates along the metal surface, by using a light beam which undergoes total internal reflection at the glass surface. At a certain angle of incidence of the light beam, surface plasmon resonance occurs and is observed as a sharp minimum of reflectance intensity. This critical angle is very sensitive to variations in the refractive index of material lying a few hundred nanometres outside the metal film. The increase in reflected light can be measured, which corresponds to the system moving out of resonance as molecules are bound to the surface.

Integrated-optical sensors have been developed recently[12] which utilize grating couplers on planar waveguides. These can be used to respond either to a change in the refractive index of a liquid sample covering the waveguide, or to the adsorption and desorption of molecules from a gaseous or liquid sample. When the incident light beam strikes the miniature diffraction grating on the waveguide, dispersion of light into the sample medium occurs. The dispersion is sensitive to the refractive index of the sample and results in a shift in the incoupling angle or the incoupled power, which can be detected by a photodetector.

Suggested further reading: Dakin & Culshaw (1988, 1989)[10,11].

Piezoelectric devices

Exploitation of acoustic techniques in biological science has received scant attention until recently. Developments in piezoelectric transducers, especially expansion of the upper ultrasonic frequency limits, has led to an increased interest in this area over the last few years. Piezoelectric crystals are mass produced cheaply for the electronics industry, an attraction for biosensor applications. The major limitations of devices such as these is, at present, the poor understanding of the acoustical characteristics of biological samples.

Piezoelectric materials are able to generate and transmit acoustic waves in a frequency-dependent manner. At a certain frequency, acoustic resonance is induced, which is a function of the crystal mass. Adsorption of molecules onto the crystal surface leads to a frequency change (ΔF) according to the Sauerbrey equation:

$$\Delta F/F = -\Delta m/A\rho t$$

where Δm = mass of adsorbed material (g), A = crystal face area (cm^2), ρ = density of adsorbed material (g cm^{-3}), and t = thickness of adsorbed layer (cm).

ΔF is highly sensitive to Δm, with values varying between 500 and 2500 Hz/μg with pg detection limits available from commercial quartz crystals. In solution, device performance is limited by response to the solution components such as electrolyte concentration, conductivity and also by non-specific adsorption. In such cases, the Sauerbrey equation is not obeyed strictly and the response is dependent on visco-elastic effects at the surface.

Surface acoustic waves can be formed by applying radiowave frequencies to one of two pairs of interdigitated electrodes (transmitter) on the surface of a piezoelectric material. An acoustic wave (Rayleigh surface wave) propagates along the surface of the crystal and is detected by a second pair of electrodes (receiver). The depth of penetration of the wave is of the order of a wavelength, and its transmission is dependent on material deposited and/or on the surface.

Suggested further reading: Clarke *et al.* (1989)[13]; Guilbault (1990)[14].

Calorimetric devices

Two transducers presently dominate research into thermal biosensors. Thermistors are temperature-sensitive, semiconductor-based sensors. Thermopiles are arrays of thermocouples designed to increase sensitivity. The former is a resistivity-based and the latter a voltage-based measurement. A third device, developed recently, utilizes a light-activated micro-fabricated resonating silicon bridge. The resonant frequency is extremely sensitive to temperature, allowing changes of millidegrees to be detected.

Measurement of temperature changes due to a biological reaction is almost universally applicable and appears straightforward. However, thermal interference is a problem which often requires elaborate and expensive equipment to overcome.

Suggested further reading: Danielsson (1989)[15].

BIOSENSOR EXAMPLES

The number of permutations of transducer and biological component is very large. In the following section, several examples of biosensors will be described that illustrate important concepts.

Oxidase enzyme electrodes based on oxygen electrodes

The first, and most widely used, enzyme in biosensor applications is glucose oxidase, which catalyses the reaction:

$$\text{glucose} + O_2 \rightarrow \text{gluconic acid} + H_2O_2$$

Glucose concentration can be monitored by measuring, for example, the fall in oxygen tension or the production of hydrogen peroxide. This can be achieved amperometrically by the use of a platinum working electrode poised at –0.6V (versus saturated calomel electrode) for the detection of oxygen or at +0.7V for the detection of hydrogen peroxide. The relatively extreme potentials required make this system prone to interference from other electroactive species. A further problem is the reliance of the reaction on the oxygen concentration in the sample.

Glucose biosensors based on the Clark oxygen electrode (as described earlier) contain an outer polycarbonate membrane which entraps glucose oxidase close to the electrode surface and excludes large molecules and cells from the external solution. An inner gas-permeable membrane restricts the species that can reach the electrode, decreasing interference. In hydrogen peroxide-detecting systems, an inner acetate membrane replaces the gas-permeable membrane, allowing the passage of hydrogen peroxide, oxygen, salts and water, but excluding glucose, ascorbic acid and some other potential interferents. Membranes, as used in this example, have proved successful in excluding most major interferents, but the approach is not ideally suited to mass production techniques.

Mediated electrochemical biosensors

In the previous example, oxygen was present as an electron acceptor. It is possible to substitute artificial electron acceptors (mediators) so that oxygen is no longer required, and hydrogen peroxide, ideally, is not produced. Using glucose oxidase as an example again, the following reaction scheme can be applied:

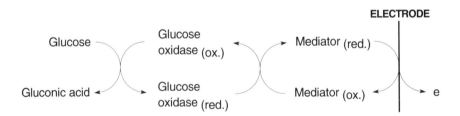

Electrochemistry occurs at the redox potential of the mediator, which can be chosen to minimize interference from other electroactive species. In addition, a successful mediator will be stable and have rapid kinetics. In particular, the reduced form must be stable in the presence of oxygen. Some of the most commonly used mediators include benzoquinone, potassium ferricyanide, tetrathiafulvalene, tetracyanoquinodimethane and the ferrocenes.

The most commercially successful biosensor to date, the ExacTech (MediSense, Cambridge, MA, U.S.A.) blood glucose meter is based on this principle[18], utilizing a ferrocene derivative as a mediator. A disposable enzyme electrode strip as used in this meter is shown in Figure 3.

Biosensors based on the fluoride ion-selective electrode

The enzyme peroxidase can be used to catalyse the cleavage of the covalent carbon–fluorine bond in certain organofluorine compounds such as 4-fluorophenol with

Figure 3. Schematic representation of the enzyme electrode strip used with the ExacTech glucose meter

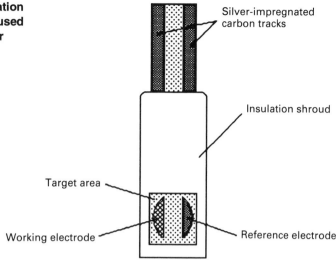

Silver-impregnated carbon tracks

Insulation shroud

Target area

Working electrode

Reference electrode

the elimination of fluoride ions. In its simplest form, hydrogen peroxide can be monitored directly by immobilizing peroxidase close to a fluoride ion-selective electrode[19] and providing 4-fluorophenol as a substrate, but the system is highly versatile. It can be coupled to any enzyme catalysed reaction in which peroxide is released. For example, glucose can be monitored by co-immobilizing glucose oxidase and peroxidase. A further advantage lies in the remarkable specificity of the fluoride electrode, which has no significant interferents at or close to neutral pH.

Electrochemical immunosensors and immunoassays

Antibodies can be labelled with electroactive molecules so that both amperometric and potentiometric measurement can be employed. There are many such systems which have been proposed and developed.

Field-effect transistors are charge-measuring devices that appear attractive for the direct detection of immunochemicals which are bound to the sensitive gate area. However, to date, attempts at achieving this have been largely unsuccessful. The reasons for this have been discussed in detail recently[20].

Enzyme-labelled antibodies can be used to catalyse a reaction releasing an electroactive product. Furthermore, a secondary enzyme can be incorporated as an amplifier which can greatly increase the apparent activity of the label. This technique has been demonstrated for the determination of prostatic acid phosphatase using an alkaline phosphatase label which catalyses the dephosphorylation of $NADP^+$ to NAD^+. This is followed by the catalytic cycling of NAD^+ to NADH in a redox cycle generating ferrocyanide (Figure 4). The rate of ferrocyanide production can be directly related to NAD^+ and hence to alkaline phosphatase concentration, and can be monitored coulometrically.

A total DNA assay system based on light-addressable potentiometric technology features disposable assay sticks which are inserted into an instrument for measurement. The system is rapid and the sensitivity is claimed to be comparable to that of conventional radiolabelling techniques. The sample is incubated with two binding

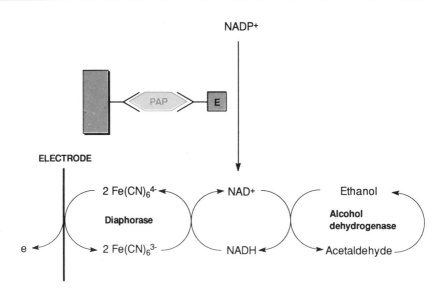

Figure 4. Schematic of electrochemical immunoassay

protein conjugates and the single-stranded binding protein is conjugated to a hapten, which makes a specific link to a capture membrane. An anti-DNA monoclonal antibody is conjugated to an enzyme, such as urease, and the labelled complex transferred to a filtration unit, where it is concentrated onto the measurement site. Unbound enzyme conjugate is washed from the membrane which is then brought into contact with the sensor. In the presence of urea the rate of change in pH can be monitored, which is proportional to the amount of DNA present (Figure 5).

Enzyme optodes

Optical oxygen and pH sensors, based on fluorescence quenching, can be used to monitor many enzyme-catalysed reactions. For example, glucose oxidase has been immobilized in a sensing layer at the end of a fibre optic light guide. The catalytic oxidation of glucose to gluconic acid lowers the local pH which can then be monitored by the quenching of a fluorescent pH-sensitive dye.

Optical immunosensors

A biological binding event can be conveniently converted into a quantitative result by use of optical techniques. Solid phase immunoassay with optical detection allows measurement of both high and low molecular mass analytes in either competitive or sandwich formats. The two optical technologies currently receiving the most attention are fluorescent evanescent wave and surface plasmon resonance.

The fluorescence capillary-fill device consists of two parallel plates, with a 0.1 mm separation. The assay capture system is covalently coupled to the baseplate, which also acts as an optical waveguide. Other reagents are deposited on the top plate, so that they dissolve upon addition of sample to the device. On completion of the immunoassay, bound fluorophore can be discriminated from free fluorophore by the evanescent optics of the baseplate waveguide. Two examples of assays that have been developed using this device are for *Rubella* antibody in serum and for human chorionic gonadotrophin, a hormone used as an indicator in pregnancy testing.

The BIAcore (Pharmacia, Uppsala, Sweden) system is a surface plasmon resonance prism which enables the researcher to monitor the formation of complexes and reveal steric and allosteric effects. The unit contains an optical detection system, autosampler and microfluidic system. The sensing interface of the BIAcore is an exchangeable sensor chip consisting of a glass support coated with a gold film and a biocompatible layer of carboxylated dextran bound to the gold film. This latter layer increases the effective surface area and sensitivity and reduces non-specific binding. A biospecific ligand is immobilized within the dextran layer, enabling interactions to be studied directly, in complex samples, without prior purification. Immobilization typically takes about 30 minutes and an average analysis between 5 and 10 minutes.

CONCLUSIONS AND FUTURE DIRECTIONS

Electrochemical techniques are likely to continue to dominate both research and commercial products in enzyme-based biosensors in the foreseeable future, and many other applications for the techniques will continue to be found. Optical techniques are extremely promising, particularly in affinity applications. The predicted advances in optical technology and materials resulting from the large amount of research in telecommunications and optical computing should increase opportunities in optical biosensors. At present, other transducer-based systems offer interesting possibilities

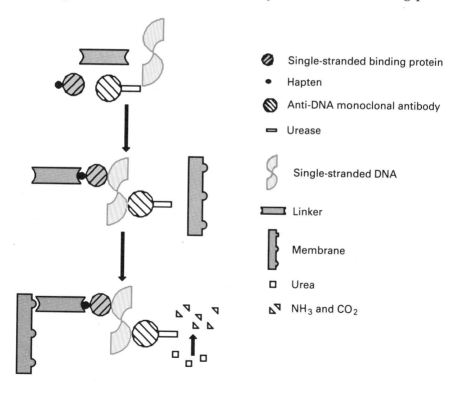

Figure 5. Schematic of light-addressable potentiometric sensor assay

for certain applications, but appear unlikely to challenge electrochemical or optical methods in the majority of situations.

Medical diagnostic requirements have been a major driving force in biosensor development: the instant analysis of clinical samples is appealing to both patient and physician. A lot of early work in this field was directed at the determination of blood glucose in insulin-dependent diabetic patients. Recently, attention has been directed towards *in vivo* sensors, which represent a particular challenge, since workers face not only the traditional problems of biosensor development, but the need for biocompatibility. In recent years, the number of biosensor applications outside of the clinical sector has been increasing. New legislation and increasing awareness of environmental issues has fuelled research into biosensors for pollutants such as pesticides, herbicides, microbial contamination and gases such as carbon monoxide. Likewise, hygiene regulations and the need for tighter quality control in an increasingly competitive market are convincing the food industry of the benefits of biosensors. Military applications are focussed on specialized requirements such as biological and chemical defence, but the development of generic technologies could have spin-offs in other applications.

REFERENCES

1. Clark, L.C., Jr. & Lyons, C. (1962) Electrode systems for monitoring in cardiovascular surgery. *Ann. N.Y. Acad. Sci.* **102**, 29–45
2. Cass, A.E.G. (1990) *Biosensors: a Practical Approach*, IRL Press, Oxford
3. Hall, E.A.H. (1990) *Biosensors*, Open University Press, Buckingham
4. Turner, A.P.F., Karube, I. & Wilson, G.S. (eds.) (1989) *Biosensors: Fundamentals and Applications*, Oxford University Press, Oxford
5. Turner, A.P.F. (1991) *Advances in Biosensors*, volume 1, JAI Press, London
6. Bard, A.J. & Faulkner, L.R. (1980) *Electrochemical Methods, Fundamentals and Applications,* John Wiley and Sons, New York
7. Wang, J. (1988) *Electroanalytical Techniques in Clinical Chemistry and Laboratory Medicine*, VCH, New York
8. Orion Research (1982) *Handbook of Electrode Technology*, Cambridge, MA
9. Bergveld, P. (1970) Development of an ion-sensitive solid-state device for neuro-physiological measurements. *IEEE Trans. Biomed. Eng.* **17**, 70–71
10. Dakin, J. & Culshaw, B. (1988) *Optical Fiber Sensors, Volume 1: Principles and Components*, Artech, London
11. Dakin, J. & Culshaw, B. (1989) *Optical Fiber Sensors, Volume 2: Principles and Components*, Artech, London
12. Tiefenthaler, K. & Lukosz, W. (1989) Sensitivity of grating couplers as integrated-optical chemical sensors. *J. Opt. Soc. Am. B* **6**, 209–220
13. Clarke, D.J., Blake-Coleman, B.C. & Calder, M.R. (1989) Principles and potential of piezo-electric transducers and acoustical techniques, in *Biosensors: Fundamentals and Applications* (Turner, A.P.F., Karube, I. & Wilson, G.S., eds.), pp. 551–571, Oxford University Press, Oxford
14. Guilbault, G.G. (1990) Piezoelectric crystal biosensors. *Am. Biotechnol. Lab.* **8(4)**, 28, 30–32
15. Danielsson, B. (1989) Theory and application of calorimetric biosensors, in *Biosensors: Fundamentals and Applications* (Turner, A.P.F., Karube, I. & Wilson, G.S., eds.), pp. 575–595, Oxford University Press, Oxford

16. Saini, S., Hall, G.F., Downs, M.E.A. & Turner, A.P.F. (1991) Organic phase enzyme electrodes. *Anal. Chim. Acta.* **249**, 1–15

17. Barker, S.A. (1989) Immobilisation of the biological components of biosensors, in *Biosensors: Fundamentals and Applications* (Turner, A.P.F., Karube, I. & Wilson, G.S., eds.), pp. 85–99, Oxford University Press, Oxford

18. Cardosi, M.F. & Turner, A.P.F. (1990) Recent advances in enzyme-based electrochemical glucose sensors. *Diabetes Annu.* **5**, 254–272

19. Pirzad, R., Newman, J.D., Dowman, A.A. & Cowell, D.C. (1989) Horseradish peroxidase assay: optimisation of carbon–fluorine bond breakage and characterisation of its kinetics using the fluoride ion-selective electrode. *Analyst* **114**, 1583–1586

20. Blackburn, G.F. (1989) Chemically sensitive field-effect transistors, in *Biosensors: Fundamentals and Applications* (Turner, A.P.F., Karube, I. & Wilson, G.S., eds.), pp. 481–530, Oxford University Press, Oxford

12

How groups of proteins titrate — a new approach

Henry B.F. Dixon

Department of Biochemistry, University of Cambridge, Tennis Court Road, Cambridge CB2 1QW, U.K.

INTRODUCTION

All discussions of the possible mechanisms for the action of an enzyme will involve specifying the states of dissociating groups in or near the active centre of the enzyme. The properties of $-COO^-$ and $-COOH$, for example, differ vastly, so it is necessary to know the ionic states of functional groups. One does not need to be primarily concerned with enzyme mechanisms to wish to know the contribution of an ionizing group to a biochemical process, and consequently how its ionic state varies with pH.

To handle this, many of us learn to cope with the titrations of monobasic acids, and textbooks deal thoroughly with the topic. Just as we learn to handle hyperbolic curves of saturation of a binding site with a ligand (Figure 1a), we learn to handle the transformed curves, such as that shown in Figure 1(b), when the concentration of ligated form is plotted against log [L] in place of [L]. This is effectively what we do when we plot the concentration of the dehydronated form of an acid against pH. [The term hydron has been recommended by IUPAC[1] as the name for H^+ when the user does not wish to imply a particular isotope.]

The complexities enter when there are two or more acidic groups in a molecule, and this is just where much of the interest enters too. Many enzymes have interacting ionizing groups at their active sites. The complexities arise because hydronation of one group affects the dissociation constant of another. So the pK of the group we are interested in changes as we titrate it, because we are simultaneously titrating other groups.

These complexities are diminished by realizing, as described below, that interacting

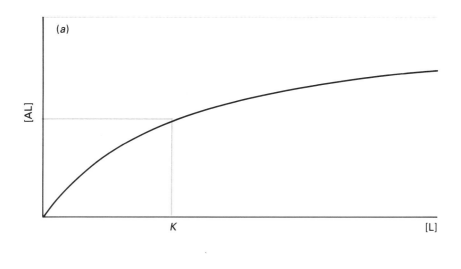

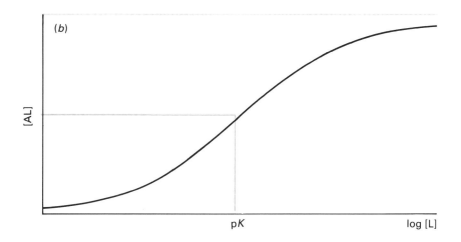

Figure 1. Single-site saturation curves
(a) The hyperbolic dependence of the concentration of the ligated form of an acceptor on the concentration of the free ligand; (b) the curve showing the dependence of the concentration of the ligated form of an acceptor on the logarithm of the concentration of free ligand. The curve is of the form of tanh x against x.

groups titrate in fractions, each fraction following a simple one-site titration curve. Further, there are simplifying relations between the pK values of the molecule and those of the groups it contains.

GROUP pK VALUES

The group HA– is likely to have a pK different in HA–BH from the one it has in

Scheme 1. The scheme of dissociation of a two-site acid

Red arrows show the dissociation of group –AH and black arrows those of group –BH. The pK of the group –AH is pK_A when the other group is in the –BH form, and becomes pK_A' when the other group is dissociated to –B⁻. Thus the group dissociation constants are defined as:

$K_A = [H^+][\ulcorner A–BH]/[HA–BH]$,
$K_B = [H^+][HA–B\urcorner]/[HA–BH]$,
$K_A' = [H^+][\ulcorner A–B\urcorner]/[HA–B\urcorner]$, and
$K_B' = [H^+][\ulcorner A–B\urcorner]/[\ulcorner A–BH]$

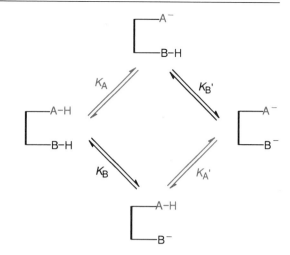

HA–B⁻ (Scheme 1). So the group HA– actually changes its pK as the substance is titrated. The change may be due to a direct influence of –BH and –B⁻ (primarily electrostatic, possibly involving hydrogen bonding), or to some change in conformation evoked by the titration of –BH. We can specify both values of the pK for HA–, although there are difficulties in determining them, and no measurement free of additional assumptions is possible[2], but checks on the necessary assumptions (e.g. Metzler et al.[3]) can make estimates reliable. Specification of all the group pK values, however, becomes complex when there are several groups.

MOLECULAR pK VALUES

A marked advance is to consider molecular pK values. For this we do not have to consider the groups separately, so the acid HA–BH can be called H₂Q, which dissociates first to HQ⁻ and then to Q²⁻ (Scheme 2). Its molecular dissociation constants K_1 and K_2 are defined as:

$$K_1 = [H^+][HQ^-]/[H_2Q]$$

and

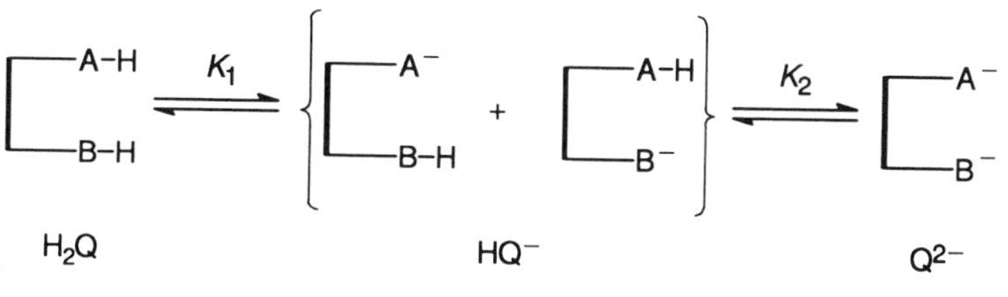

Scheme 2. The molecular dissociation constants of a two-site acid

From the equations given in Scheme 1, K_1, defined as $[H^+][HQ^-]/[H_2Q]$, is equal to $K_A + K_B$, so is itself a constant. Likewise, K_2, defined as $[H^+][Q^{2-}]/[HQ^-]$, is equal to $K_A' \cdot K_B'/(K_A' + K_B')$, so is also constant. p$K_1$ and pK_2 are known as the molecular constants.

$$K_2 = [H^+][Q^{2-}]/[HQ^-]$$

but already we must be careful about the meaning of the symbols. $[HQ^-]$ represents the sum of two concentrations, those of the two species that differ according to which of the two groups has lost H^+, i.e.:

$$[HQ^-] = [^-A{-}BH] + [HA{-}B^-]$$

Molecular pK values show many advantages over group constants. First, there are far fewer of them. Second, they can be unambiguously assigned. Third, by not distinguishing between the tautomers $^-A{-}BH$ and $HA{-}B^-$, they correspond to our lack of knowledge; any measurement we make of how some property changes with pH may tell us how we lose H^+ as the pH is raised, and so how we pass from H_2Q to HQ^- and on to Q^{2-}, but it gives no direct information on the relative amounts of the tautomers. We do not need the separate values of the concentrations of these species, since the ratio $[^-A{-}BH]/[HA{-}B^-]$ is independent of pH; hence the pH-dependence of any property is fully described by assigning a value of the property we are measuring to the total $[HQ^-]$ as well as to $[H_2Q]$ and $[Q^{2-}]$.

There is, however, something artificial about molecular constants. Suppose that two molecules of an acid with a single dissociating group are combined to form a dibasic acid, and that the original pK of the monobasic acid is pK_0. Suppose that the groups in the dibasic acid are so far apart that they do not affect each other. Hence identical and independent groups result, and combination of the two molecules of acid has not changed the titration curve, that of a monobasic acid of pK_0. The definitions of pK_1 and pK_2 nevertheless force us to assign to the acid two different molecular values, $pK_1 = pK_0 - 0.3$, and $pK_2 = pK_0 + 0.3$. The 0.3 arises as $\log(2)$, since the first H^+ to be lost has two identical sites to yield it, but only one free for it to return to, whereas the converse is true for the second, whose dissociation constant is fourfold lower, i.e. its pK is 0.6 higher. All equations of pH-dependence, including the bell-shaped ones representing the pH-dependence of $[HQ^-]$, are easily represented in terms of the molecular constants, as Michaelis[4] did in 1914, but we have lost touch with the groups that give the compound its acidity.

Figure 2 (facing page). Demonstration that fractions of a single group in a dibasic acid can possess different pK values[5]

(a) The concentrations of the three forms of H_2Q are plotted against pH. The intersections marked are pK_1 and pK_2 because when pH = pK_1 then $[H_2Q] = [HQ^-]$, and when pH = pK_2 then $[HQ^-] = [Q^{2-}]$. (b) The red and black curves show the individual forms of HQ^-, forms differing according to which group is hydronated. The intersections marked show the group pK values, since when pH = pK_A, then $[HA{-}BH] = [^-A{-}BH]$, when pH = pK_B, then $[HA{-}BH] = [HA{-}B^-]$, when pH = pK_A', then $[HA{-}B^-] = [^-A{-}B^-]$, and when pH = pK_B', then $[^-A{-}BH] = [^-A{-}B^-]$ (see Scheme 1). (c) The red curve shows the sum of the concentrations of the forms in which the group HA– is dissociated. This demonstrates how fractions of this group exhibit different pK values. The black curve similarly shows the sum of the concentrations of the forms in which the group HB– is dissociated. *Note:* To make this figure clear the groups HA– and HB– have been made to differ little in pK (by 0.26), so that an appreciable fraction of each is dissociated in the form HQ^-. Likewise they are assumed to interact strongly, so that dehydronation of either raises the pK of the other by 2.55. This places the titration and molecular pK values of the molecule 3.2 units apart.

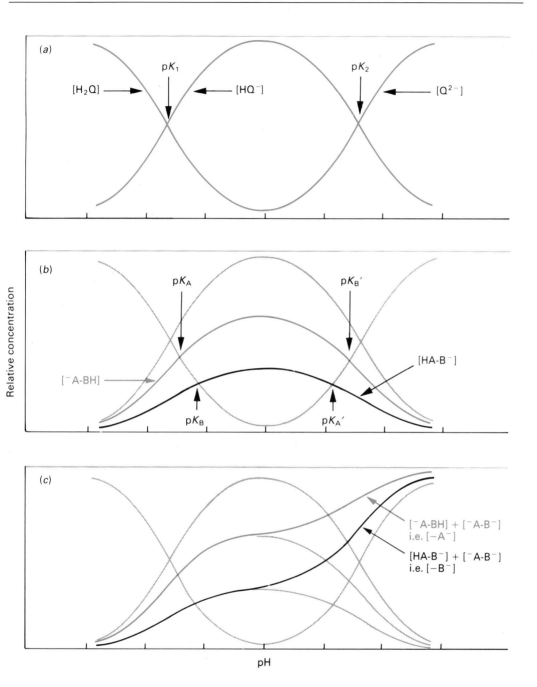

ANOTHER LOOK AT GROUPS — THEY TITRATE IN FRACTIONS

Although group pK values are inconvenient, because we cannot specify how a group switches among its values as other groups change their ionic forms, there is another approach to the behaviour of individual groups, as shown in Figure 2. Using molecular values, we first plot the pH-dependence of $[H_2Q]$, $[HQ^-]$ and $[Q^{2-}]$ (Figure 2a). I have chosen a value of pK_1 well below pK_2 because this makes the figure clearer. It ensures that the first hydron dissociates almost completely as the pH is raised before dissociation of the second becomes appreciable. It is also a reasonable assignment, because the extra negative charge on Q^{2-} can give it a higher affinity for H^+ than HQ^- possesses, i.e. $pK_2 >> pK_1$. Then we draw in the pH-dependences of $[^-A–BH]$ and $[HA–B^-]$, each a fixed fraction of $[HQ^-]$ (Figure 2b). We make them comparable in height, i.e. making the groups HA– and HB– of comparable acidity, again to make the figure clear. To study the behaviour of the group HA–, we need to plot (Figure 2c) the sum $[^-A–BH]$ + $[^-A–B^-]$, the forms in which HA– is dissociated. The resulting red line shows that a fraction of the group dissociates with one pK and the rest with another. When we[5] pointed this out in 1973, we mistakenly thought that it was only an approximation. This was because we tried to fit the fractions of the groups to the molecular pK values, but it turns out (see below) that it is a precise statement when the correct type of pK value is used. The black line similarly shows how the group –BH titrates in fractions. These have the same two pK values, but with a different fraction belonging to each of them.

A DIFFERENT APPROACH — TITRATION CONSTANTS

In 1926 Simms[6], working in the Rockefeller Institute for Medical Research (now the Rockefeller University), New York, pointed out that the titration curve of a multivalent acid was identical to that of a mixture of hypothetical univalent acids, each of the same concentration as the multivalent acid. He called the dissociation constants of these hypothetical acids the "titration constants" of the real acid, and symbolized them as G (his expressions for them are given below). They approached the molecular constants if these were well separated. Figure 3 gives the same system as Figure 2, and shows the curve of hydron release against pH; it is the sum of the two one-site curves marked.

For some time little use was made of these constants. In 1975 I[7] pointed out that the "Michaelis functions" (also called alpha functions, as Michaelis[4] used the symbol α for them), the expressions in terms of K_1 and K_2 for the relative concentrations $[H_2Q]_r$, $[HQ^-]_r$ and $[Q^{2-}]_r$ (the subscript 'r' signifying that each is expressed as a fraction of their sum) are respectively the products of concentrations of forms of hypothetical acids HX and HY, such that:

$[H_2Q]_r = [HX]_r \cdot [HY]_r$

$[HQ^-]_r = [HX]_r \cdot [Y^-]_r + [X^-]_r \cdot [HY]_r$

$\qquad$ = a constant times $[HX]_r \cdot [Y^-]_r$

$\qquad$ = a different constant times $[X^-]_r \cdot [HY]_r$

$[Q^{2-}]_r = [X^-]_r \cdot [Y^-]_r$

Thus one-site curves whose constants are the titration constants are factors of the

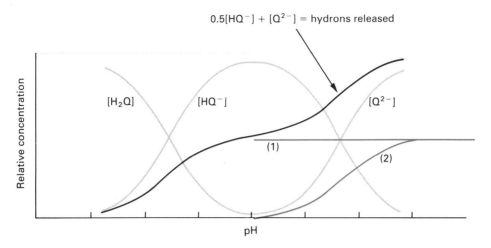

Figure 3. The titration curve of a dibasic acid is the sum of two one-site curves
The black curve of hydron release against pH for the acid shown in Figure 2 is clearly the sum of the two one-site curves (1) and (2) shown in red. This is less obvious when the pK values are closer together.

Michaelis functions. (Unfortunately I did not then know the work of Simms, so did not realize that the dissociation constants of the hypothetical acids HX and HY were exactly his titration constants.)

These equations are trivially obvious when the groups are independent, since the dissociation constants of HX and HY are then the group constants (since the titration constants are defined as constants for groups on separate molecules, and hence independent). For such a system the fraction of the acid hydronated on both groups is the fraction hydronated on one multiplied by the fraction hydronated on the other.

This factorization led to an application of the titration constants, when we[8] were able to make use of it in considering enzymes, such as a dehydrogenase catalysing the reaction:

$$NAD^+ \; + \; H-\underset{R}{\overset{R}{C}}-O-H \; \rightleftharpoons \; NAD-H \; + \; \underset{R}{\overset{R}{C}}=O \; + \; H^+$$

in which H^+ is a product in one direction and a reactant in the other. If such an enzyme gives a typical bell-shaped dependence of rate (strictly, the specificity constant, i.e. $k_{cat.}/K_m$) on pH, then the pK that characterizes the alkaline side of the bell for the direction of hydron uptake must be the same as that characterizing the acid side of the bell for the reverse direction (Figure 4). This was easy to demonstrate when all the groups in the enzyme and substrate were independent; treating such independent pK values as titration constants was one of the two ways that allowed the conclusion to be generalized to enzymes and substrates that possess interacting groups.

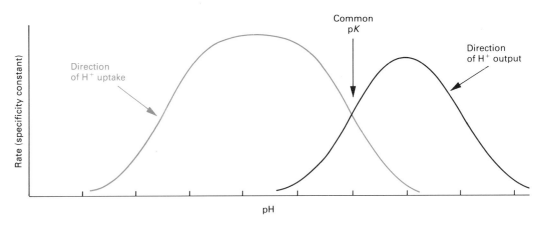

Figure 4. A possible pH-dependence of the rate in both directions of an enzyme-catalysed reaction in which H⁺ is a product/reactant

If such a reaction shows a typical bell-shaped pH-dependence, the same pK characterizes the alkaline side of the bell for the direction of H⁺ uptake and the acid side of the bell for the reverse direction[8]. This corresponds with the fact that the K_m of a substrate is usually the same as this substance's K_i as an inhibitor of the reverse reaction. Thus as the pH drops across the common pK, the rate in the direction of H⁺ uptake (red curve) rises because H⁺ is a substrate, and its K_m, expressed in the pK, is identical to its K_i as a product inhibiting the reverse reaction (black curve). The simplest reason why the activity might show the bell-shaped dependence on pH is that one group in the enzyme needs to be in its hydronated form, and another dehydronated, for the enzyme to be active.

An important use[9] of titration constants is the comparison between the titrations of an unmodified protein and of one in which one ionizing group is specifically masked. The method and its derivatives give evidence of the ionizing groups that have their pK values affected by the ionic state of the group modified.

PROPERTIES FOLLOW ONE-SITE TITRATION CURVES

In discussions after a Ph.D. oral examination on how the work should be presented and followed up, three of us, candidate, supervisor and examiner, had to write down some algebra. We then realized that we were looking at pH-dependences in an unfamiliar, but simple, way. We also saw that what we could prove for two interacting groups should be generalized to any number of groups, so we collaborated with a mathematician for this[10].

The first result was to show that it was not only the titration curve, i.e. the pH-dependence of hydron release, that could be represented as the sum of one-site curves, as Simms had shown, but that the pH-dependence of *any* property of a dissociating substance was similarly the sum of one-site curves of the form of Figure 1(*b*). These component curves were characterized by the *same* constants as for hydron release, i.e. the titration constants of the substance. Thus the dependence of *any* property on pH can be expressed by allotting fractions of its change with pH to the various hypothetical acids, HX, HY, etc. Whereas for release of H⁺, one hydron per molecule is released with each pK, the fractional changes of property associated with the various pK values may vary widely.

Since the pH-dependence of a property is made up of the sum of curves of the form of Figure 1(*b*), it is valid, as often done, to infer a pK value from one such inflection in a curve of pH-dependence. Those who do so, however, may often be unaware that the value obtained is a titration value, and this may be neither a molecular nor a group value if interacting groups are involved in determining the property.

THE TITRATION OF GROUPS

Since the pH-dependence of any property can be expressed as the sum of one-site curves, and one such property is the degree of dehydronation of a particular group, it follows that the titration of each group is expressed as the sum of one-site curves. Thus the obviously doubly-sigmoid titration of group HA– in Figure 2(*c*) is simply the sum of two one-site curves with the titration values of the molecule.

The example of cysteine can be used to bring out the features of this approach. Table 1 shows how its groups partition among its three titration pK values. The values chosen are based on the assignments of Benesch & Benesch[11], since, as mentioned above, direct measurement of the group pK values is not possible. The rows show how the groups partition among the three titration pK values. Dissociation of H^+ from the –SH and $-NH_3^+$ groups behaves like that from –AH and –BH of Figure 2(*c*). Hence a fraction of each of these groups titrates with each of the titration pK values of 8.36 and 10.53. In general, a fraction of each group titrates with each titration pK, though this fraction may be negligible, as it is for the carboxy group with the two highest pK values. Since each group has one H^+ to dissociate, these fractions add up to unity, as shown by the summation at the ends of the rows.

Table 1. The partitioning of groups among the titration pK values of cysteine.
Each group releases one hydron on titration, a release spread over the three pK values of the molecule as shown in the rows of the Table. One hydron is also released for each titration pK, and this is made up of contributions from the groups, as shown in the columns of the Table.

Group	Titration pK values			Sum for group
	1.9	8.36	10.53	
Carboxy, –COOH	1.00	0	0	1
Mercapto, –SH	0	0.68	0.32	1
Ammonio, $-NH_3^+$	0	0.32	0.68	1
Sum for pK	1	1	1	

Note: Assignment of the group pK values requires measurement of the molecular pK values, together with attribution of a value to the tautomeric ratio $[^-OOC-CH(-NH_3^+)-CH_2-S^-]/[^-OOC-CH(-NH_2)-CH_2-SH]$. This assignment was made[11] on the basis of u.v. absorbance, together with the assumption that $-S^-$ has the same absorption coefficient irrespective of the hydonation state of the amino group. The assumption may not be precisely true, especially since hydronation of the amino group changes the wavelength of maximum titration slightly, and the affinity of $-S^-$ for H^+ by 30-fold. The titration pK values differ negligibly from the molecular values, since the molecular values differ so greatly from each other.

Table 1 also illustrates a second point of the analysis. This is the summation of the columns to unity. One H^+ is released per molecule for each pK because the fractions of the different groups that titrate with this pK add up to unity. A third point is why a group may be fully associated with a single pK, and therefore negligibly with any of the others. This could be because it was independent, i.e. its affinity for H^+ unaffected by the hydronation state of any of the other, e.g. because it was too distant from them. Clearly this is not true of the carboxy group of cysteine, which is close to the other two ionizing groups. Here it is because its dehydronation is virtually completed as the pH is raised before that of the others is appreciable. Hence a group shows a single pK provided it titrates in a different pH range from any whose ionization state affects it.

The value of this method of regarding groups as titrating in fractions lies in the picture it gives us of the titrations. We are familiar with the type of curve shown in Figure 1(*b*), so we can handle the idea of applying it to each of the fractions of a group, or to the different fractional changes of any other property. Since this approach is mathematically identical to the use of molecular constants, nothing can be calculated by this approach that could not be calculated without it; its utility lies in helping us to picture ionizations.

ENZYME MECHANISMS

Several enzymes contain two ionizing groups close together in their active centres, e.g. the two histidine residues of ribonuclease A, recently reviewed in this series[12], and the two aspartic residues of pepsin. One of these groups may need to be hydronated and the other dehydronated for activity, the former acting as a general acid and the latter as either a general base or a nucleophile. Enzymologists often speculate what group is responsible for a pK observed in kinetics; a recent example is for the two aspartic residues of pepsin: 'which carboxyl group has the lower pK value?'[13]. The more realistic question (although extremely difficult to answer!) is how much

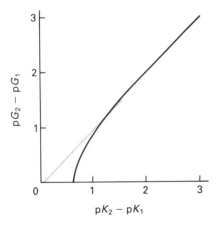

Figure 5. The difference in the titration pK values of a dibasic acid as a function of the difference between its molecular pK values

The titration values have the same mean as the molecular values; their difference is $2 \cdot \log \{[m \pm (m^2 - 4)^{1/2}]/2\}$ where the molecular values differ by $2 \cdot \log m$.

each contributes to any measured pK value. If the groups are close together, as they were here, and have similar constants, here likely since both are carboxy groups in similar environments, then we must expect appreciable amounts of both the HA–B⁻ and ⁻A–BH tautomers to co-exist.

IONIC FORMS OF GROUPS

Awareness that groups titrate in fractions can simplify arguments on their ionic forms. An example is the discussion of the state of the phospho group of pyridoxal phosphate in aspartate aminotransferase. Its ^{31}P-n.m.r. spectrum shows a pK of about 6.5 (the value varies with isoenzyme and with salt concentration) in the pH-dependence of the chemical shift of the phosphorus atom. When this was first observed, it was attributed to titration between $-PO_3H^-$ and $-PO_3^{2-}$; it was soon realized, however, that the change in shift was too small for this conversion. It was therefore concluded that the group was in the $-PO_3^{2-}$ form throughout the pH range 5–8, and that an ionization elsewhere in the protein affected the phosphorus, e.g. by changing O–P–O bond angles. But although the arguments were convincing that complete titration of $-PO_3H^-$ to $-PO_3^{2-}$ would give a much larger change in chemical shift, no one argued whether *a fraction* of the group might not possess the observed pK. Application of the present treatment[14] showed that a small fraction of the group might indeed be in the $-PO_3H^-$ form, and change the shift of the phosphorus atom as it titrated to $-PO_3^{2-}$. If this were true it would not change the essence of the previous conclusions: certainly the major part of the group is dianionic over the whole pH range 5–8, and the ionization that affects the phosphorus environment must be largely elsewhere. A neighbouring –XH group could stabilize the bound $-PO_3^{2-}$ form even at pH 5, possibly by hydrogen bonding; then the form at this pH would be an equilibrium mixture of mainly $-PO_2^-$–O⁻ ⋯ H–X– and some $-PO_2^-$–O–H ⋯ ⁻X–. So, as the pH was raised to give $-PO_2^-$–O⁻ ⋯ X–, part of the phospho group would titrate with the pK of 6.5, just as part of –AH titrates with the upper pK in Figure 2; most of the ionization with pK 6.5 would be that of –XH, not of the phospho group. In fact, the same conclusions on the ionic forms had been reached by Metzler & Metzler[15] on the basis of estimating tautomeric equilibria from u.v. spectra.

TYING UP SOME LOOSE ENDS

What are the titration constants?

Of course Simms[6] gave equations relating the molecular constants, K_1, K_2, etc, with the titration constants G_1, G_2, etc. When there are more than two groups, the equations become fairly fierce, and explicit solutions for each titration constant are complicated. For two groups, where $K_1 = G_1 + G_2$, and $K_1K_2 = G_1G_2$, the explicit solutions for the titration values are:

$$G_2, G_1 = \{K_1 \pm [K_1(K_1 - 4K_2)^{\frac{1}{2}}\}/2$$

or, in pK values, if m and pK^* are defined[16] by equating the molecular constants, pK_2 and pK_1, with $pK^* \pm \log m$, then the titration values pG_2 and pG_1 are:

$$pK^* \pm \log \{[m \pm (m^2 - 4)^{\frac{1}{2}}]/2\}$$

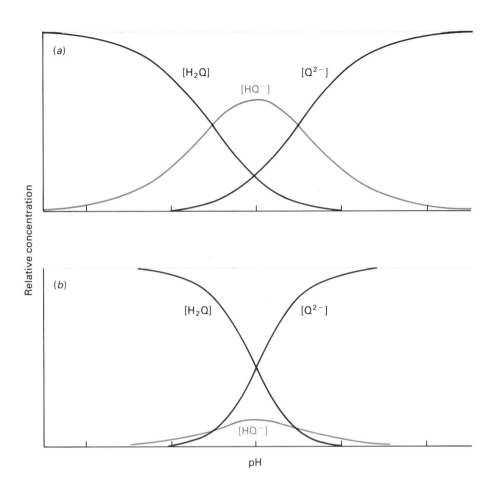

Figure 6. pH-dependences of concentrations of the ionic forms of a dibasic acid at different co-operativities of H$^+$-binding

(a) Negative co-operativity, i.e. $pK_2 > pK_1 + 0.6$. Curves are shown for $pK_2 - pK_1 = 1$. They provide a further example to that of Figure 2(a), for which $pK_2 - pK_1 = 3.2$. Each curve is the sum of one-site curves, and the product of one-site curves. (b) Positive co-operativity, i.e. $pK_2 < pK_1 + 0.6$. Curves are shown for $pK_2 - pK_1 = -1$. Since real values do not exist for the titration constants under these conditions, these curves are not the sums and products of one-site curves. Because of the positive co-operativity of H$^+$ binding, the concentration of HQ$^-$ is always relatively low.

(and thus have the same mean, pK^*). When $pK_2 - pK_1 = 0.6$, then $pG_2 - pG_1 = 0$, i.e. the titration constants are equal. As the difference $pK_2 - pK_1$ rises, $pG_2 - pG_1$ rises faster (Figure 5), so that the titration constants become effectively equal to the molecular constants when these differ by over two units; the titration values differ from the molecular values by 0.01 when the molecular values differ by 1.7, and by only 0.002 when the molecular values differ by 2.4.

So although the titration constants form a series such that pK_1 is lower than pG_1, and pK_n, the highest molecular pK, is higher than pG_n, the highest titration value, each titration value is likely to be quite close to the corresponding molecular value if these are reasonably separated.

Are the titration constants real?

I will only begin to answer this question, and only in its mathematical sense. We have seen above that two identical and independent groups give molecular pK values that differ by 0.6, and two identical titration values; effectively there is only one group that titrates, although two copies of it per molecule. Two factors can raise $pK_2 - pK_1$ above 0.6: (1) any difference between the groups in their affinity for H^+, and (2) interaction between the groups, commonly dictated by electrostatics, so that loss of H^+ from one group makes loss of one from the second more difficult. In quantitative terms[5]:

$$K_1/K_2 = (K_A/K_A')(2 + K_A/K_B + K_B/K_A)$$

where K_A and K_B are the dissociation constants of groups $-AH$ and $-BH$ respectively when the other is hydronated, and K_A' is that of $-AH$ when $-BH$ is dehydronated. It is conceivable, however, that loss of one H^+ may actually facilitate loss of a second, because, for example, it leads to a conformation change; this situation is described as positive co-operativity of hydronation. Hence a dibasic acid could have $K_1 < 4 K_2$, i.e. $pK_2 < pK_1 + 0.6$. Then the expressions shown above give imaginary numbers as the values to the titration constants, and the titration will not be the sum of two one-site curves.

Figure 6 illustrates this. In Figure 6(a), $pK_2 - pK_1 = 1$, and each of the curves shown is the sum of two one-site curves (as well as being the product of two one-site curves!); the titration pK values differ by 0.90. When, however, pK_2 is below $pK_1 + 0.6$, no real titration values exist. So the curves shown in Figure 6(b), where $pK_2 - pK_1 = -1$, are not the sums or products of one-site curves. It may be noted how low the curve for [HQ⁻] is; this reflects the positive co-operativity of binding, so that addition of H^+ to Q^{2-} increases the affinity for H^+, and HQ^- easily passes on to H_2Q.

For a multivalent acid it turns out[10] that all the titration values are real if each molecular constant is at least four times the value of that for the next hydron to dissociate, i.e. the pK values differ by 0.6. But if this is not true, the analysis may collapse, although not completely[10], in that many of the one-site curves will still be present, and an unreal pair can be replaced by a two-site curve showing positive co-operativity, i.e. a curve like one of those in Figure 6(b). Such a curve therefore replaces two one-site curves in the sum that gives the pH-dependence of any property, e.g. the degree of dissociation of one of the groups in the molecule.

FEWER CONSTANTS TO ASSIGN

One of the advantages of assigning fractions of groups to the n titration pK values is that we have fewer assignments to make than if we assign group pK values. Each group has a different value according to the ionization state of each of the other $n - 1$ groups, i.e. it has 2^{n-1} values. The molecule therefore has $n \cdot 2^{n-1}$ group pK values, which works out at over 5000 (Table 2) where there are 10 groups. Not all of these are independently assignable. If, for example, we have a two- group molecule, HA–BH, it has four values, the two values for HA– in both HA–BH and HA–B⁻ (red

Table 2. Numbers of group dissociation constants and of assignments of groups among titration constants

Groups...	n	1	2	3	4	5	...	10
Group pK values	$n \cdot 2^{n-1}$	1	4	12	32	80		5120
Independently assignable group pK values	$2^n - 1$	1	3	7	15	31		1023
Allotments of titration pK values	$n^2 - n + 1$	1	3	7	13	21		91

in Scheme 1), and the two for –BH in HA–BH and in ⁻A–BH (black in Scheme 1). But since the equilibrium constant for forming ⁻A–B⁻ from HA–BH must be the same whether the route is via ⁻A–BH or via HA–B⁻, fixing any three of the four group constants fixes the fourth. There are therefore much fewer independently assignable group values, as Table 2 also shows.

If we assign titration pK values and the groups to them, we have fewer assignments to make[10]. The first n assignments are the titration values themselves. We then assign a group to these values. For the first $n - 1$ titration constants, we have a value to assign. After these assignments, the rest of the group is assigned to the nth constant. Such assignments have to be repeated for the rest of the first $n - 1$ groups, making $(n - 1)^2$ assignments in all. The last group does not have to be assigned, as it must make up the remaining fraction to make each of the titration pK values have one H^+ per molecule. Adding these $(n - 1)^2$ assignments to the original n assignments of titration values makes a total of $n^2 - n + 1$ assignments. As Table 2 shows, this is fewer than the number of independently assignable group values if $n > 3$, over 10-fold fewer when n is 10.

There has to be some loss in making fewer assignments. Its nature may be seen in considering an acid $H_{10}Q$, and particularly the form in which it has lost five hydrons, namely H_5Q^{5-}. The assignments of groups to the various pK values specifies how much of each of the 10 groups is dehydronated in H_5Q^{5-}, but it does not specify how this fraction partitions between the many forms according to which other groups are dissociated. Specifying how much of each group is dissociated in H_5Q^{5-} is only 10 specifications; but there are 252 tautomers whose relative amounts could be specified.

CONCLUSIONS

● 1. An acid with n ionizing groups is known to have the same titration curve as an equimolar mixture of n hypothetical monobasic acids, whose pK values are called the "titration" pK values of the real acid.

● 2. A group in the molecule titrates, not with one pK, but as the sum of fractions. Each fraction shows one of the titration pK values of the acid. Thus each group partitions among the pK values (Table 1).

- 3. Although each group in a molecule with n ionizing groups possesses 2^{n-1} group pK values, e.g. 512 for $n = 10$, its degree of dissociation is completely described by the n fractions that titrate with the n titration constants of the molecule.

- 4. One H^+ per molecule titrates with each pK; it is made up of fractions derived from the n different groups. This is similar to the one H^+ that dissociates from each group; it is made up of fractions, each exhibiting one of the n different pK values.

- 5. The pH-dependence of all properties of the molecule can be represented as the sum of one-site titration curves. Each of these curves possesses one of the n titration pK values.

- 6. It is therefore important to remember, when a pK is found in the pH-dependence of a property such as enzyme activity, that more than one group may contribute to it, and that these groups may contribute to different degrees.

- 7. This description helps in considering the mechanisms and pH-dependences of enzymes.

I thank many colleagues for helpful discussion and for teaching me about multiple dissociations, especially Professor K. Brocklehurst, Dr T.K. Carne, Dr S.D. Clarke, Dr A. Cornish-Bowden, Dr R.M. Dixon, Dr P.A. Evans, Dr A.J. Kirby, Professor D.E. Metzler, Rev. Dr A.R. Peacocke, Dr F.J.C. Rossotti, Dr G.A. Smith, Professor K.F. Tipton, Dr M. Vas, and Dr S.G. Waley.

REFERENCES

1. Bunnett, J.F. & Jones, R.A.Y. (1988) Names for hydrogen atoms, ions, and groups, and for reactions involving them. *Pure Appl. Chem.* **60**, 1115–1116
2. Dixon, H.B.F. (1976) The unreliability of estimates of group dissociation constants. *Biochem. J.* **153**, 627–629
3. Metzler, D.E., Harris, C.M., Johnson, R.J., Siano, D.B. & Thomson, J.A. (1973) Spectra of 3-hydroxypyridines. Band-shape analysis and evaluation of tautomeric equilibria. *Biochemistry* **12**, 5377–5392
4. Michaelis, L. (1914) in *Die Wasserstoffionenkonzentration*, pp. 30–39, Springer, Berlin [English translation (1926) pp. 55–69, Ballière, Tyndall and Cox, London]
5. Dixon, H.B.F. & Tipton, K.F. (1973) Negatively co-operative ligand binding. *Biochem. J.* **133**, 837–842
6. Simms, H.S. (1926) Dissociation of polyvalent substances. I. Relation of constants to titration data. *J. Am. Chem. Soc.* **48**, 1239–1250
7. Dixon, H.B.F. (1975) Factorization of the Michaelis functions. *Biochem. J.* **151**, 271–274
8. Dixon, H.B.F., Brocklehurst, K. & Tipton, K.F. (1987) pH–Activity curves for enzyme-catalysed reactions in which the hydron is a product or reactant. *Biochem. J.* **248**, 573–578
9. Parsons, S.M. & Raftery, M.A. (1972) Ionization behaviour of the catalytic carboxyls of lysozyme: effects of ionic strength. *Biochemistry* **11**, 1623–1629
10. Dixon, H.B.F., Clarke, S.D., Smith, G.A. & Carne, T.K. (1991) The origin of multiply sigmoid curves of pH-dependence. *Biochem. J.* **278**, 279–284
11. Benesch, R.E. & Benesch, R. (1955) The acid strength of the –SH group of cysteine and related compounds. *J. Am. Chem. Soc.* **77**, 5877–5881

12. Parés, X., Nogués, M.V., de Llorens, R. & Cuchillo, C.M. (1991) Structure and function of ribonuclease A binding subsites. *Essays Biochem.* **26**, 89–103

13. Sielecki, A.R., Fedorov, A.A., Boodhoo, A., Andreeva, N.S. & James, M.N.G. (1990) Molecular and crystal structures of monoclinic porcine pepsin refined at 1.8 Å resolution. *J. Mol. Biol.* **214**, 143–170

14. Dixon, H.B.F. (1991) Part of the phospho group of pyridoxal phosphate may titrate over the pH-range 5–8 in aspartate aminotransferase. *Biochem. J.* **280**, 832–833

15. Metzler, C.M. & Metzler, D.E. (1987) Quantitative description of absorption spectra of a pyridoxal-phosphate-containing enzyme using lognormal distribution curves. *Anal. Biochem.* **166**, 313–327

16. Dixon, H.B.F. (1988) Relations between the dissociation constants of dibasic acids. *Biochem. J.* **253**, 911–913

Correction

An error occurred in the chapter on ribonuclease A binding sites by Parés *et al.* in volume 26 of *Essays in Biochemistry*. Figure 10 on page 99 was shown as having been reproduced, with permission, from *Science*; in fact, this figure was an original by the authors. Figure 8 on page 96 was reproduced from *Science*. Our apologies to all concerned!

Subject index